Practical and Clear Graduate Statistics in Excel

The Excel Statistical Master

(that'll be you!)

By Mark Harmon

mark@ExcelMasterSeries.com
www.ExcelMasterSeries.com
ISBN: 978-1-937159-13-9

Reader Testimonials

"I bought Mark Harmon's Excel Master Series manual as a reference for a graduate course on statistics that I was taking as part of an MBA program at the University of Delaware. I purchased the materials about halfway through the course and wish I had known about this manual from the start of the class!

Mark has done a great job in writing complex statistical concepts in an easy to understand format that makes grasping them both easy to understand and to use.

With the help of Mark's book, and some diligent studying, I received an A in my stats course.

Thanks Mark!"

Chris Veale
Newark, Delaware

"The Excel Statistical Master really saved me in my graduate statistics class last semester. The book that was used in the class really did not give the practical down to earth instructions I needed to apply my statistical knowledge to excel.

In this quide I was able to find helpful step by step instructions (and pictures) that walked me through creating ANOVA's, Hypothesis testing, and so much more.

I highly recommend this book for graduate students and managers who are looking to maximize the power of Excel in their daily operational activities."

Christopher M. Walden
Jacksonville Beach, Florida

"Mark, Thanks for your invaluable material. I have used it in my current business research methods for my bachelor students. It has enabled me to abandon SPSS for the first time!

I have also given your link to the last MBA class I taught. They told me that your book was of great help to understand my course!"

Professor Emmanuel Fragnière
HEG Geneva and University of Bath

"After years of searching for a simplified statistics book, I found the Excel Statistical Master.

Unlike the indecipherable jargon in the countless books I have wasted money on, the language in this book is plain and easy to understand.

This is the best $40 I have ever spent."

Mahdi Raghfar
New York, New York

"I am a medical student at Semmelweis University and the Excel Statistical Master helped me so much with passing my midterms and my semifinal exam. There is no way I would have passed without it.

Even though I went to all of the classes and consultations, it was the Excel Statistical Master that taught me all of the basic concepts for the different tests we used.

Each test is explained in different steps and how you performed it on Excel. Illustrations and screenshots make it easy to follow, even for those like me that never had used Excel before.

I highly recommend Excel Statistical Master for all medical students. It's worth every dollar. And I have to say that the communication with the seller have been the best! If I had questions about statistics problems, he more than gladly answered them. It's so easy and saved my from hours with reading!

Thanks a lot!"

Annette Myhre
Medical Student
Semmelweis University
Budapest, Hungary

"I am taking evening courses to get my degree in business administration at the University of Applied Sciences in Friedberg, Germany. During the day I am a sales manager in a production facility.

For my bachelors thesis, I am performing a comprehensive statistical analysis of repair costs at the facility that I work.

After searching for days on Google for the right framework to solve this problem, I finally found the solution. The Excel Statistical Master has allowed me to find exactly the right distributions and showed me how to create some excellent graphs. The explanations and videos in the manual are excellent, even for a non-native English speaker!

Thanks Mark!"

Frank A. Mathias
Facility Management Major
Bachelor of Business Administration
University of Applied Sciences
Friedberg, Germany

"Whenever I evaluate a book on statistics, I always look at the table of contents to check both the topics covered and the examples.

Not only does the author cover all of the techniques you're likely to need in a graduate program, he also goes into substantial detail on when to use each technique.

I also found that his case studies and focused examples, all of which are listed in the detailed table of contents, were on point and good learning tools that will help learners know when and how to apply the techniques.

His explanations, particularly on when (and how) to use two-tailed tests for hypothesis testing and the Poisson and exponential distributions, were clear and will help anyone learn how to implement these techniques.

In particular, this book will help anyone without a substantial background in math, such as many political science M.A. students, to learn and apply the concepts.

Practical and Clear Graduate Statistics in Excel is very well done and well worth your money."

Curtis D. Frye
Author of
Microsoft Excel 2010 Plain & Simple

"I really like the Excel Statistical Master. It is incredibly useful. The explanations and videos in the manual are excellent.

It has really made my work with statistics a LOT easier. I'm really glad that I came across the manual.

If you're a student of business statistics, this e-manual is worth WAY more it's priced. I will use your manual as a reference for my Econometrics MBA course that I will be teaching this summer."

Dr. Yan Qin
Adjunct Assistant Professor of New York University
Adjunct Assistant Professor of Keller Graduate School of Management
Co-Director - Nankai-Grossman Center for Health Economics and Medical Insurance
New York, New York

"Faced with a seemingly intractable spreadsheet assignment for my online Operational Research course at a UK university and very little time, <u>my purchase of this book could not have come at a better time.</u>

With clear steps to follow, mastering the Statistical 'matter' and being able to apply them with Microsoft Excel assisted a great deal and broke down the 'hard nuts' in my spreadsheet assignment quickly.

<u>I recommend this book if you are afraid of 'statistics'. It'll definitely drive those fears far away.</u>"

Toyin Lamikanra
Online Masters Program in Operational Research
University of Strathclyde

"I bought the Excel Statistical Master to help me in my statistics class. I must say, it was unbelievably useful. Not only did I master statistics in Excel, but <u>the e-manual actually did a much better job of explaining statistics than my text book did.</u>

That e-manual made my statistics class much easier to understand, and I am now able to do all of that stuff in Excel, easily ! It is a GREAT book ! <u>If you're a student or business manager wanting to learn statistics, this is easiest, fastest way to do it.</u>

Thank you again for everything."

Tiran Ovsepyan
North Hollywood, CA

"At first I was quite skeptical of this book and the attendant Microsoft Excel spreadsheets. Everyone touts how easy their product is and how you really "need" their product to be successful.

For one of the few times I can remember, that claim is fulfilled with this product.

I have been using Excel since the product first came out on the Macintosh. But the Excel Master series has shown me how really limited my knowledge was and how much there is yet to learn.

The Excel Master series is certainly an enjoyable and well constructed method of learning statistics. The quality of the author's effort to reduce a relatively boring subject to easily understood examples and text is evident throughout.

I highly recommend this series to anyone interested in the application of statistics to everyday problems using to most powerful computer environment available."

John G. Black
Warriors Mark, Pennsylvania

"I'm a PhD consultant in the area of "user experience". I help companies make their web sites user friendly. When I conduct a usability test on a web site design, I need to determine if one design is better than another. So I need to conduct a t-test. At other times, I conduct focus groups that help my client determine which

of, say, 10 product designs is better than the others. So I need to conduct non-parametric ANOVAs on their rating responses.

All this goes to say, I need to deal with data rapidly, and in a manner that I can send to my client. Does my client have any statistical software? No, they don't have SyStat or MiniTab or SPSS, or any other packaged stats program. However, my clients DO have Excel.

With Excel, they can open my spreadsheet and see the data. While they are looking at the data (and perhaps running their own descriptive statistics like means, etc.) they can also see my statistical analysis I conducted with the Excel Statistical Master. They can see my charts, too.

So, you can imagine the benefit this gives to my client. It's "one-stop shopping". Data, analysis, and charts all appear in one file. I like being able to send one file that works with Microsoft Office. My client has Microsoft Office. Now they have everything they wanted: stats, data, and charts."

John Sorflaten, PhD Certified Usability Analyst (CUA)
Certified Professional Ergonomist (CPE)
Sr. Usability Engineer www.saic.com
Columbia, Maryland

"I'm a Statistician with more than 25 years of field work, at the same time I'm probably the top MS Excel expert in the Middle East.

Excel now is a Universal tool and an absolute must for those wishing to understand practical Statistics and applying Statistics to a very large percentage of real life situations.

All of the above facts require a clear understanding of the statistical features and capabilities of Excel. What is needed is a guide that takes you to the point DIRECTLY.

As an Instructor Mr. Harmon's excellent work saves me considerable time, and should help all students. And as a quick and direct revision tool for those applying Statistics in Business, Insurance, Medical Decision making and many more situations.

I can tell You that this Masterpiece is Very Very Very useful and would save you considerable time if you teach, but If you are a Student...Please Pray for the Author... This is a GIFT from GOD.

Keep it Up please."

Abdul Basit AL-Mahmood
Senior Consultant
ExcelTech
Kingdom of Bahrain

"We just started building statistical Excel spreadsheets for our direct mail and online marketing campaigns, I purchased Excel Statistical Master to help fill in some of the blanks.

Little did I know, this book has everything I could ever want to know about business statistics. Easy to follow and written so even a child could understand some of the most complex statistical theories.

Thanks Mark!"

Brandon Congleton
Marketing Director
www.WorldPrinting.com
Clearwater, Florida

"The Excel Statistical Master is a real life saver in my forensic accounting practice. Until I found the package I was struggling with some of the "how to" aspects of statistics..no more. The videos are an extra bonus that really help!

I am very pleased!!!"

Glenn Forrest
Seattle, WA

"As an engineering major in college, I never took statistics.

Now that I am a practicing engineer, statistics is an important part of my job - from reviewing test data to designing experiments and performing ANOVA.

The Excel Statistical Master series has helped me get up to speed faster than any traditional textbook.

I also found that my technical reports are more polished and professional as a result of my study of this material. The lessons are organized into logical groupings of topics and are just the right size for self-study or if you are using it as a course supplement.

Once I had mastered some of the basic topics I was able to skip around and study those topics I needed for my work. The ability to follow along with the exercises in Excel and then test out new scenarios is a real bonus and a great way to gain a feel for some of the concepts.

Since most of my work is done with Excel anyway, the exercises have served as templates for my job.

I highly recommend this series to anyone who works with statistics and also to the professional who desires to have a nicely organized reference for performing statistical calculations in Excel."

Chris Bronnenberg
Los Angeles, California

"Mark, I am quite impressed with the Excel Statistical Master.

As a research practitioner, I've used the easy-to-understand document to help work through some pretty daunting data sets.

As a professor of research, I will suggest the eManual as a supplement to my students.

***I think the strength of the eManual is the straightforward explanations of complex procedures** in a software platform that is readily available. SPSS and SAS should shudder at the competition.*

Great job!"

Tait J. Martin, Ph.D.
President and Chief Insight Officer
The iNSiGHT Cooperative
Tallahassee, Florida

"The Excel Statistical Master eManual is a wonderful product for anyone who needs to apply a variety of statistical tools and does not have the time or background to develop those tools themselves.

The Excel Statistical Master is easy to use, comprehensive, and powerful.

Congratulations on an excellent product!"

Cliff Sather
Bennington, New England

"Going through Excel Statistical Master has helped me in filling the gaps which most of the professional ignore while building models from scratch.

There are many books in market available in market but the Excel Statistical Master explains everything in a simple way and how to use Excel to solve real life problems. Every topic in Excel Statistical Master is self explanatory and I would recommend freshmen as well as professionals to go through Excel Statistical Master.

Thanks Mark."

Ashutosh Gupta
Sr. Financial Analyst/Team Lead Mortgage Industry Advisory Corporation
Bengaluru, India

"My first encounter with the EXCEL Statistical Master came when writing a report for a customer when the axiom "an un-used tool becomes rusty" presented itself. I knew what analysis and presentation tool I wanted, but using EXCEL became cumbersome and frustrating.

A quick on-line check revealed the EXCEL Statistical Master which proved to be unequivocal, easy to follow, and complete.

Only after completing the report did I notice the description on the front page "Clear And Simple yet Thorough." Clearly, my experience supported substantiated this claim. Seldom does one encounter such truth in advertising.

Last weekend the manual was very useful in helping my grandson, who just started a statistics class in high school, help understand, envision, and define his semester project as to which data to gather and how to analyze and present that data.

I have recommended the Excel Statistical Master to customers and my contractors."

Pat Goodman
SSL Consulting
Morgan Hill, California

"I am an IT Consultant who gives a Data Analysis for Decision Making workshop to various private and public sector organizations. This workshop consists of many statistical methods.

I often use the Excel Statistical Master in my workshop to demonstrate procedures, give usable examples and frequently, learn new procedures myself.

I find it easy to use, clear and succinct. It should wipe out the fear of statistics from those who have a block against it."

Akram Najjar
Director
InfoConsult
IT Consulting
Beirut, Lebanon

Table of Contents

Overall Course Modules

Chapter 1 - Histograms, Charting, and Descriptive Statistics

Creating a Chart

Problem: Chart the Dog vs. Cat Population data:

	A	B	C
7	Pet Population in U. S.		
8	Year	Dogs	Cats
9	1948	40,619	14,974
10	1949	40,803	15,580
11	1950	41,129	16,285
12	1951	40,831	17,000
13	1952	40,712	17,593
14	1953	41,334	17,957
15	1954	41,496	17,492
16	1955	41,749	18,266
17	1956	42,645	19,456
18	1957	42,625	19,591
19	1958	42,833	20,093
20	1959	43,053	20,455
21	1960	43,563	20,689
22	1961	43,907	21,608
23	1962	43,589	21,758
24	1963	44,025	22,134
25	1964	44,397	22,734
26	1965	44,837	23,351
27	1966	44,698	24,043
28	1967	45,086	25,003
29	1968	45,671	25,642
30	1969	46,081	26,770
31	1970	46,842	27,954
32	1971	47,627	28,810

	A	B	C
30	1969	46,081	26,770
31	1970	46,842	27,954
32	1971	47,627	28,810
33	1972	48,542	29,580
34	1973	49,389	30,148
35	1974	50,862	31,491
36	1975	51,213	32,972
37	1976	51,753	34,214
38	1977	52,784	35,399
39	1978	54,077	37,323
40	1979	55,349	38,959
41	1980	56,225	40,747
42	1981	56,860	41,866
43	1982	57,461	42,952
44	1983	58,105	44,255
45	1984	59,250	44,994
46	1985	59,949	46,740
47	1986	61,126	47,852
48	1987	61,899	49,085
49	1988	62,423	50,436
50	1989	63,375	51,996
51	1990	64,805	52,925
52	1991	65,149	53,328
53	1992	65,767	54,356
54	1993	66,329	54,982
55	1994	66,788	56,322
56	1995	67,516	56,871
57	1996	67,434	57,503
58	1997	68,884	58,788
59	1998	69,547	59,583
60	1999	70,295	60,718

To create a chart for this data, locate the charting tool at:

Insert / Chart

1) Select the following chart type:

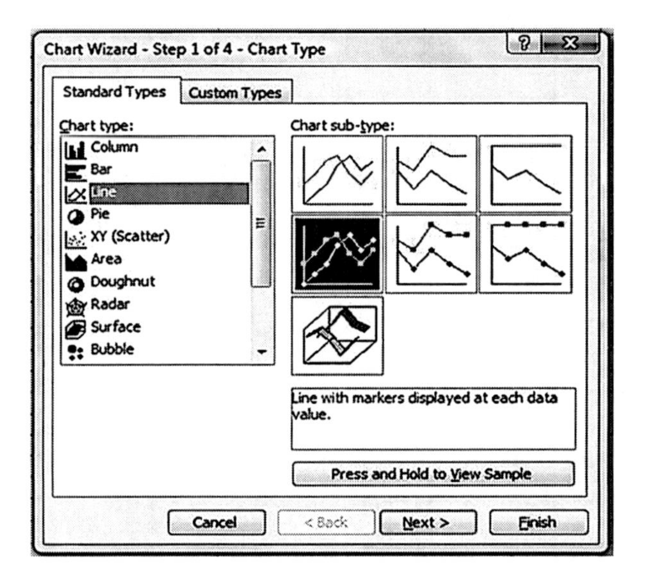

2) Select the "Columns" because the data is in columns. Input the data range by highlighting the darker right columns of Dog & Cat data, including the column labels - Dogs & Cats

After the data columns are highlighted, a preliminary graph will appear in the dialogue box as shown below:

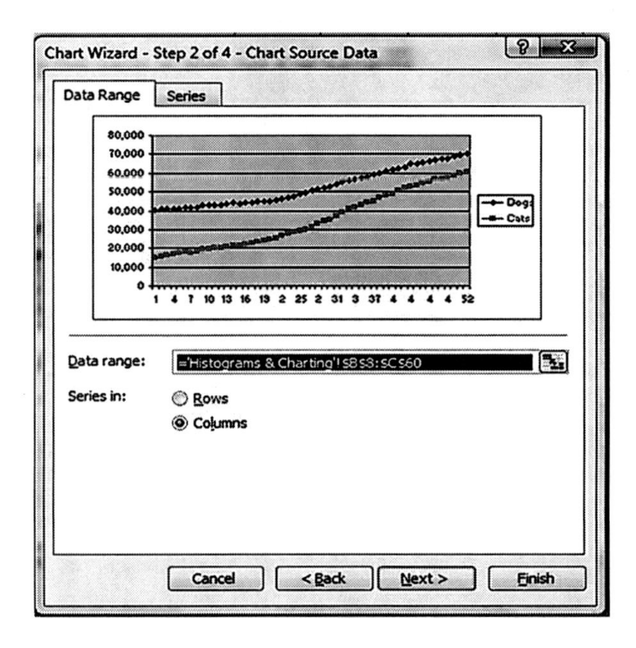

Click on the Series Tab. For Category (X) axis labels, select the date column of data as is lighter-highlighted left Year column. The dialogue will then appear as below:

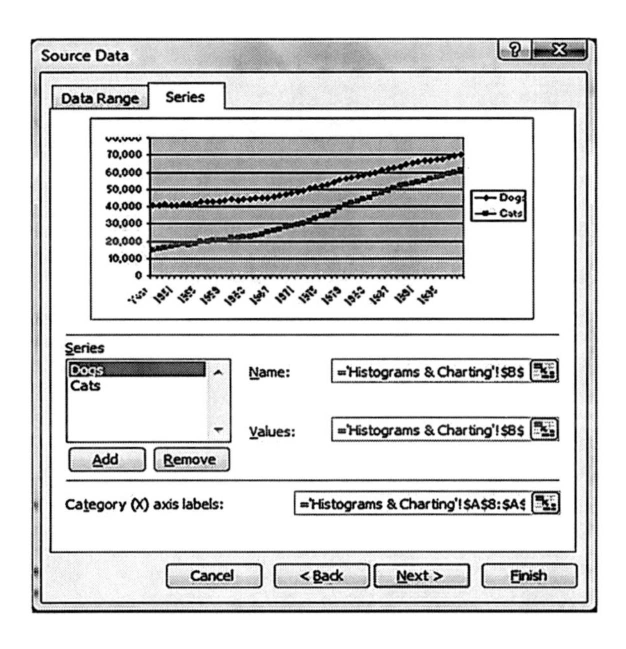

3) Type in the Chart Title, Category (X) axis, and Value (Y) axis as is shown below:

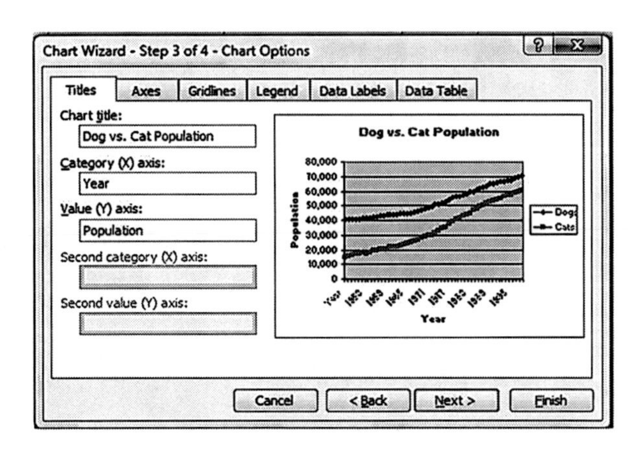

4) Select Finish and the following chart appears:

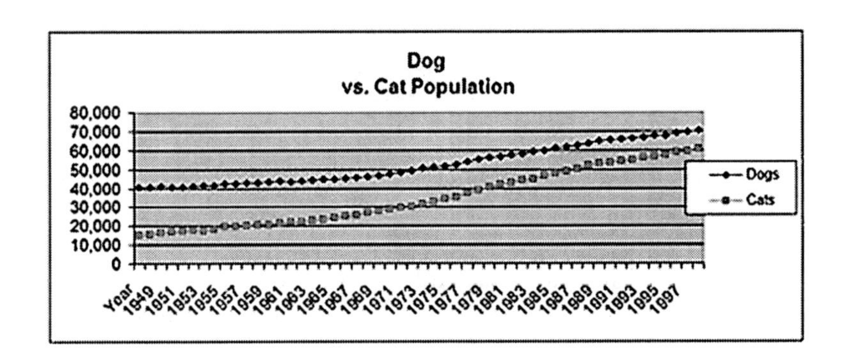

Creating Descriptive Statistics

Problem: Create Descriptive Statistics for the Dog vs. Cat Data:

Descriptive Statistics can be found at:

Tools / Data Analysis / Descriptive Statistics
(in Excel 2010 -> Data tab)

The following dialogue box appears:

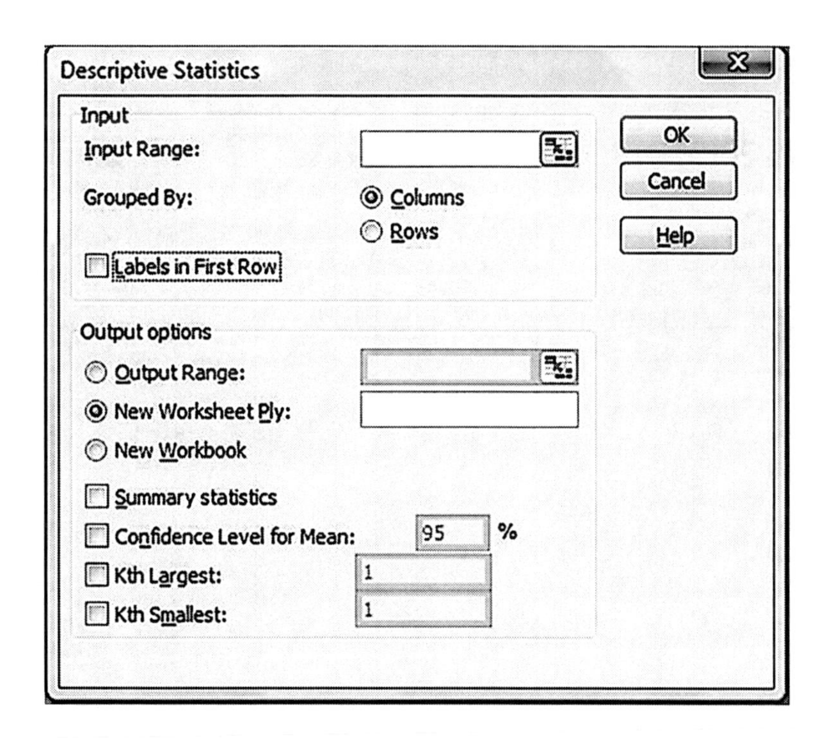

For the Input Range, highlight only the data columns and labels, just as are highlighted darker (the Dogs & Cats columns on the right) above.

Select "Labels" if the column labels were highlighted along with the data, as was done in this example.

In this case, only Summary Statistics was selected.

Clicking OK will produce the following output:

Descriptive Statistics ----> Tools / Data Analysis / Descriptive Statistics

Highlight only the data columns and labels, just as are highlighted here in yellow

Dogs		*Cats*	
Mean	52371.30769	Mean	34646.5962
Standard Error	1362.939367	Standard Error	2051.74486
Median	50125.5	Median	30819.5
Mode	#N/A	Mode	#N/A
Standard Deviation	9828.295549	Standard Deviation	14795.3426
Sample Variance	96595393.39	Sample Variance	218902163
Kurtosis	-1.329434402	Kurtosis	-1.36515367
Skewness	0.41216714	Skewness	0.3414429
Range	29676	Range	45744
Minimum	40619	Minimum	14974
Maximum	70295	Maximum	60718
Sum	2723308	Sum	1801623
Count	52	Count	52

Creating a Histogram

Instructional Video

Go to
http://www.youtube.com/watch?v=BEV3AgpEoTo
to View a
Video From Excel Master Series
About How To Create
a Histogram and Pareto Chart
in Excel

(Is Your Internet Connection and Sound Turned On?)

Problem: Create a Histogram for the Student Population Data Below:

State University	Student Population
Alabama	25,000
Alaska	10,000
Arizona	22,000
Arkansas	4,000
California	260,000
Colorado	40,000
Connecticut	45,000
Delaware	12,000
District of Columbia	24,000
Florida	145,000
Georgia	46,000
Hawaii	15,000
Idaho	20,000
Illinois	55,000
Indiana	65,000
Iowa	20,000
Kansas	45,000
Kentucky	60,000
Louisiana	10,000
Maine	40,000
Maryland	80,000
Massachusetts	175,000
Michigan	95,000
Minnesota	55,000
Mississippi	12,000
Missouri	49,000
Montana	2,000
Nebraska	80,000
Nevada	30,000
New Hampshire	100,000

Maryland	80,000
Massachusetts	175,000
Michigan	95,000
Minnesota	55,000
Mississippi	12,000
Missouri	49,000
Montana	2,000
Nebraska	80,000
Nevada	30,000
New Hampshire	100,000
New Jersey	85,000
New Mexico	35,000
New York	145,000
North Carolina	65,000
North Dakota	30,000
Ohio	95,000
Oklahoma	75,000
Oregon	25,000
Pennsylvania	175,000
Rhode Island	20,000
South Carolina	30,000
South Dakota	15,000
Tennessee	45,000
Texas	155,000
Utah	30,000
Vermont	80,000
Virginia	95,000
Washington	50,000
West Virginia	25,000
Wisconsin	80,000
Wyoming	45,000

Locate the Histogram tool at:

Tools / Data Analysis / Histogram
(in Excel 2010 -> Data tab)

The following dialogue box will appear:

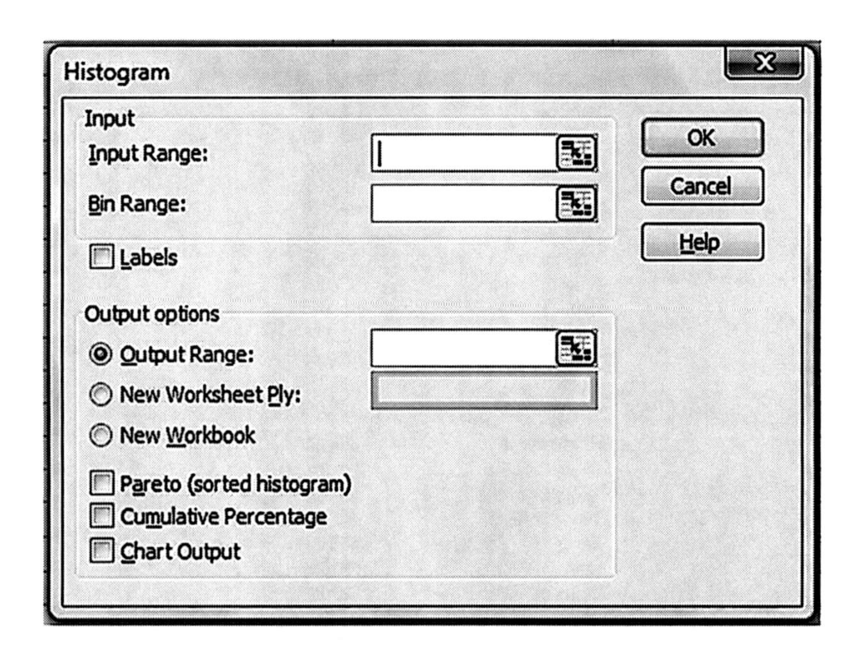

1) Create the Bins that you want the data arranged in, as done below:

When asked to select Bin Range, select the column with the lower limits of each bin range, as is highlighted darker below:

Bin Range You Must Create - Histogram's Input is the Highlighted Lower Limit Column		
Interval	More than ..	But not more than..
1	1000	30000
2	30000	50000
3	50000	70000
4	70000	100000
5	100000	125000
6	125000	150000
7	150000	200000
8	200000	240000

2) Select the Input Range of data. In this case, it would be the column of Student Population data highlighted darker previously. Select that data column.

3) Select "Labels" (if they were highlighted with the Input data). Select an Output range (select the cell that will be the upper left corner of the data output. Check "Pareto" and "Chart Output."

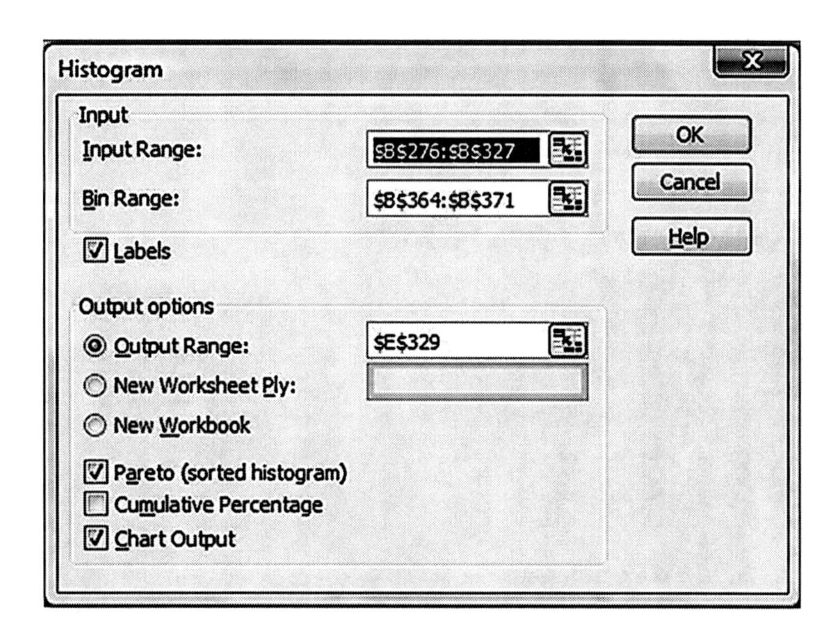

Select OK and the following output will be generated:

Histogram Output

1000	Frequency	Cumulative %	1000	Frequency	Cumulative %
30000	20	39.22%	30000	20	39.22%
50000	10	58.82%	50000	10	58.82%
70000	5	68.63%	100000	10	78.43%
100000	10	88.24%	70000	5	88.24%
125000	0	88.24%	200000	3	94.12%
150000	2	92.16%	150000	2	98.04%
200000	3	90.04%	More	1	100.00%
More	1	100.00%	125000	0	100.00%

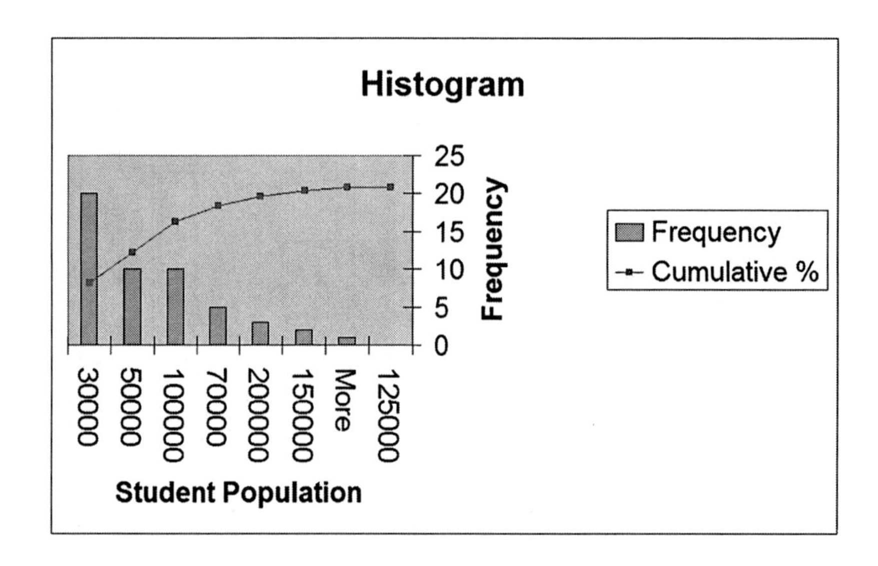

Chapter 2 - Combinations and Permutations

Combinations and Permutations are smaller groupings of objects often selected from a larger population. Objects in Combinations are selected simultaneously from the population. Objects in Permutations are selected sequentially from the population.

Basic Explanation of Combinations and Permutations

The concepts of combinations and permutations are closely related. A typical problem will ask how many combination or permutation groups containing x number of objects can be obtained from a larger population containing n objects.

Difference Between Combinations and Permutations

The major difference between a combination and a permutation is **when** the elements of the group are chosen:

Combinations - Elements are picked simultaneously, all at once.

Permutations - Elements are picked sequentially, one after another.

If there is any order to the arrangements, it is a Permutation

Combination Formulas

The Number of Combinations = $_n\mathbf{P}_x$ **/ x! = n! / [x! * (n - x) !]**
of n different objects taken x at a time simultaneously

Excel Functions Used When Calculating Combinations

COMBIN (n, x)

= Number of Combinations of n Objects Taken x at a Time Simultaneously
----> This is a Math & Trig category Function, not a Statistical category function like most of the functions used in this course.

FACT (n) = n!
----> This is a Math & Trig Function.

For example, the number of Combinations of 9 objects taken 4 at a time Simultaneously =

= $_n\mathbf{P}_x$ **/ x!**
= $_9\mathbf{P}_4$**/ 4!**

= n! / [x! * (n - x)!]
= 9! / [4! * (9 - 4)!]

= FACT(n) / (FACT(x) * FACT(n - x))
= FACT(9) / (FACT(4) * FACT(5))
= 126

= COMBIN (n,x)
= COMBIN (9,4)
= 126

Permutation Formulas

The Number of Permutations = ${}_nP_x$ = n! / (n - x) !

of n different objects taken x at a time **sequentially**

Excel Functions Used When Calculating Permutations

PERMUT (n, x) = Number of Combinations of n Objects Taken x at a Time Sequentially

----> This is a Math & Trig category Function, not a Statistical category function like most of the functions used in this course.

FACT (n) = n!

For example, the number of Permutations of 9 objects taken 4 at a time Sequentially =

= ${}_nP_x$ = ${}_9P_4$ = n! / (n - x)!
= 9! / (9 - 4)!

= FACT(n) / FACT(n - x)
= FACT(9) / FACT(5)
= 3,024

= PERMUT (n,x)
= PERMUT (9,4)
= 3,024

Combination Problems:

Problem 1: Combinations of Investment Proposals

Problem: A company is evaluating 6 investment proposals. If the company selects 3 of the proposals simultaneously, how many different groups of three investment proposals can be selected?

This is a combination problem because the three investment proposals are selected **simultaneously**.

n = 6 = total number of investment proposals available for inclusion in each combination group

x = 3 = number of investment proposals that will be **simultaneously** selected to fill each combination group

The number of combinations of n = 6 different investment proposals selected x = 3 at a time **simultaneously** equals:

${}_nP_x$ / x! = ${}_6P_3$ / 3! = **COMBIN (n,x)** = COMBIN (6,3) = 20

Problem 2: Combinations of Newly Opened Offices

Problem: A consultancy wants to open 4 offices in 10 Northern states. Each new office will be in a different state. The offices will open all at the same time. How many different ways can these four offices be situated among the 10 possible Northern states?

This is a combination problem because the four different states are to selected as locations simultaneously.

n = 10 = total number of states available for inclusion in each combination group

x = 4 = number of states that will **simultaneously** be selected to fill each combination group

The number of combinations of n = 10 different states available to selected at x = 4 at a time **simultaneously** equals:

$_{n}P\mathbf{x}$ / x! = $_{10}P_{4}$ / 4! = **COMBIN (n,x)** = COMBIN (10,4) = 210

Problem 3: Multiple Combinations of Newly Opened Offices

Problem: A consultancy wants to open 4 offices in 10 Northern states, 3 offices in 9 Southern states, and 2 offices in 8 Eastern states. Each new office will be in a different state and all offices will be opened at the same time. How many different combinations does the company have to evaluate?

Total number of combinations =

= COMBIN(10,4) * COMBIN(9,3) * COMBIN(8,2) = 493,920

= 210 * 84 * 28 = 493,920

Problem 4: Combinations of Committees

Problem: From a group of 10 men and 8 women, a committee is formed. The committee will have 3 men and 4 women. How many different ways can this committee of 3 men and 4 women be formed from the overall group of 10 men and 8 women? All committee members are picked at the same time.

This is combination problem because all committee members are picked **at the same time**. The problem asks how many ways can all possible combinations of 3 out of 10 men be combined with all possible combinations of 4 out of 8 women?

Total number of combinations =

= (All possible combinations of men) * (All possible combinations of women)

= COMBIN(10,3) * COMBIN(8,4) = 8,400

= 120 * 70 = 8,400

Problem 5: Combinations of Sub-Groups

Problem: How many ways can a group of 12 people be divided into one group of 7 and another group of 5?

This is a combination problem because all members of any one group can be picked **simultaneously.**

One way to solve the problem would be to determine the total number of 7-person combinations that can be formed from 10 people and then multiply that number by the number of 5-person combinations that can be formed from the remaining 5 people.

Total number of combinations =

= (total number of combination of 7 out of 12) * (total combinations of 5 out of remaining 5)

= COMBIN(12,7) * COMBIN(5,5) = 792

= 792 * 1 = 792

Permutation Problems:

Problem 6: Permutations of Delivery Routes

Problem: A milkman makes 7 deliveries on his route. How many different sequences can he make to complete all 7 stops?

This is a permutation problem because the stops are done **sequentially**.

n = 7 = total number of objects initially available for inclusion in each permutation group

x = 7 = number of objects that will **sequentially** fill the permutation group

The number of permutations of n = 7 different stops taken x = 7 at a time **sequentially** equals:

$_nP_x = {_7P_7}$ = **PERMUT(n,x)** = PERMUT(7,7) = 5,040

Problem 7: Permutations of Seating Arrangements

Problem: How many ways can 5 people be seated on a sofa if only 3 seats are available and the 3 seats are filled sequentially by the available 5 people?

This is a permutation problem because the elements of the permutation group are filled **sequentially**.

n = 5 = total number of objects initially available for inclusion in each permutation group
x = 3 = number of objects that will **sequentially** fill the permutation group

The number of permutations of n = 5 different people seated x = 3 at a time **sequentially** equals:

$_nP_x = {_5P_3}$ = **PERMUT(n,x)** = PERMUT(5,3) = 60

Problem 8: Permutations of Executive Groups

Problem: A group of 9 people needs to appoint 1 person to be group president, another person to be group vice president, and a third person to be group treasurer. If the group first votes for the president, then votes for the vice president, and finally votes for the treasurer, how many different executive groups can be created from the original 9 people?

This is a permutation problem because the elements of the permutation group are filled up **sequentially**.

n = 9 = total number of objects initially available for inclusion in each permutation group

x = 3 = number of objects that will **sequentially** fill the permutation group

The number of permutations of n = 9 different people elected x = 3 at a time **sequentially** equals:

${}_nP_x = {}_9P_3$ = **PERMUT(n,x)** = PERMUT(9,3) = 504

Problem 9: Permutations of Book Arrangements

Problem: How many ways can 3 books be placed next to each other on a shelf one at a time?

This is a permutation problem because the elements of the permutation group are filled **sequentially**.

n = 3 = total number of initially objects available for inclusion in each permutation group

x = 3 = number of objects that will sequentially fill the permutation group

The number of permutations of n = 3 different books placed x = 3 at a time **sequentially** equals:

$_nP_x$ = $_3P_3$ = **PERMUT(n,x)** = PERMUT(3,3) = 6

Problem 10: Permutations of Letter Groups

Problem: From the following six letters: A, B, C, D, E, F, how many groups of 3 letters can be created if none of the letters from the original 6 are repeated in any group?

This is a permutation problem because none of the letters can be repeated. When the first letter of one of the permutation groups is chosen, there are only five remaining letters to choose from. Thus, the elements of the permutation group are filled sequentially.

n = 6 = total number of objects initially available for inclusion in each permutation group

x = 3 = number of objects that will **sequentially** fill the permutation group
The number of permutations of n = 6 different letters placed x = 3 at a time **sequentially** equally:

$_nP_x$ = $_6P_3$ = **PERMUT(n,x)** = PERMUT(6,3) = 120

Hand Calculation of Combination and Permutation Problems

Go To
http://excelmasterseries.com/Excel_Statistical_Master/Combinations-Permutations.php

To View How To Solve Combination and Permutation Problems By Hand (No Excel)

(Is Your Internet Connection Turned On ?)

You'll Quickly See Why You Always Want To Use Excel To Solve Statistical Problems !

Chapter 3 - Correlation and Covariance Analysis

Correlation and Covariance describe relationships between different variables. Both Correlation and Covariance describe whether variables move together in the same direction, move in opposite directions, or don't move in any related way at all.

Basic Explanation of Correlation and Covariance

Correlation and Covariance are very similar ways of describing relationships between two variables. Correlation is a more well-known concept and more widely used. It will therefore be covered in the first half of this course module. Covariance will be covered in the second half.

The Difference Between Correlation and Covariance

A common question in statistics: What is the difference between correlation (the Spearman correlation) and covariance analysis.

First, let's discuss the commonality between correlation and covariance. Correlation and Covariance both describe relationships between 2 variables.

The Spearman Correlation is considered to be a standardized form of covariance.

Here are the differences between the Spearman Correlation and Covariance Analysis:

Correlation values fall within the range of -1 to +1. Covariance values can be outside of that range.

In the covariance calculation, covariance values depend on the units of measure of X and Y. Covariance values of data sets using different sets of measure are not comparable. The Spearman correlation coefficient is not influenced by the units of measure when calculating correlation. The Spearman correlation can be used to compare the similarities of multiple data sets that use different units of measure or scale. The Spearman correlation solves the units-of-measure problem by normalizing the covariance to the product of the standard deviations of all variables being compared. The dimensions or units of measure are then taken out of the equation.

The Spearman correlation tells you how close or far two variables are from being independent from each other. It must be remembered that high Spearman correlation interpretation does not imply causality between the two variables. The covariance calculation and covariance analysis tells you how much two variables tend to change together.

Correlation Analysis

Positive Correlation vs. Negative Correlation

Positive Correlation

If two variables are "positively correlated," they move in the same direction. When one goes up, the other goes up as well. Two variables that are positively correlated have a correlation coefficient that is between 0 and +1. **The closer the correlation coefficient is to +1, the more exactly the two variables move together**.. A correlation coefficient between two variables of exactly +1.00 means that both variables move in lock-step with each other. A correlation coefficient between two variables of 0 indicates that there is no relationship between the movement of one variable and movement of the other variable.

Negative Correlation

If two variables are "negatively correlated," they move in opposite directions. When one goes up, the other goes down. When one variable goes down, the other goes up. Two variables that are "negatively correlated" have a correlation coefficient that is between -1 and 0. **The closer the correlation coefficient is to -1, the more exactly the two variables move in opposite directions**. A correlation coefficient between two variables of exactly -1.00 means that both variables move lock-step with each other in opposite directions. A correlation coefficient between two variables of 0 indicates that there is no relationship between the movement of one variable and movement of the other variable.

Calculation of Correlation Coefficient

The correlation also describes how linear a relationship is between two variables. The Correlation Coefficient can have values between -1 and +1. Below is the formula for calculating the Correlation Coefficient. Excel does such a great job in calculating correlation and covariance that it is not necessary to memorize the formulas of covariance and correlation, but here they are, along with examples worked out in Excel:

Correlation of variables x and y from a known population = **ρ** ("rho")

ρ = (Covariance of x and y) / (Standard Deviation of x * Standard Deviation of y)

ρ = Population Correlation Coefficient = σ_{xy} / (σ_x * σ_y)

(Covariance will be explained later in this module)

Correlation of variables x and y randomly sampled from an unknown population = **r**
(This is the normal situation)

r = (Sample Covariance of x and y) / (Sample Standard Deviation of x * Sample Standard Deviation of y)

r = Sample Correlation Coefficient = S_{xy} / (S_x * S_y)

Excel Functions Used When Calculating Correlation Coefficient

CORREL (Highlighted block of cells of 2 variables)
= **r** = Sample Correlation between two variables x and y

The Excel Statistical function **CORREL** calculates the correlation between 2 variables. The only inputs needed in the **CORREL** formula are specifying the locations of the blocks of cells for each variable. This is done by highlighting each block of cells of values to be correlated after the function is inserted. **CORREL** is one of the statistical functions. **CORREL** can be used to calculate the correlation between only two variables. The Excel Data Analysis tool Correlation, which is discussed after **CORREL**, can calculate correlations between multiple variables simultaneously.

Here is an example of **CORREL** in use:

Problem 1: Calculating Correlation Between Two Variables

Problem: Calculate the correlation between variables x and y based upon the 6 pairs of x-y data given below:

x	y
1	2
3	6
6	7
8	9
5	6
4	5

The Correlation of variables x and y is $\mathbf{r_{xy}}$:

$\mathbf{r_{xy}}$ **= CORREL (darker highlighted block, lighter highlighted block)**
= 0.941884

Correlation (Highlighted block of multiple variables)

If you wish to calculate the correlation between more than 2 variables, you would use the Correlation data analysis tool. This can be found at:

Tools / Data Analysis / Correlation (in Excel 2010 -> Data tab)

Problem 2: Calculating Correlation Between Multiple Variables

Problem: Determine the correlation between all of the variables below:

x	y	z	a
1	2	10	24
3	6	9	45
6	7	8	56
8	9	7	46
5	6	6	67
4	5	5	23

The only input for the Correlation function is to highlight all of the sample data and the row or column containing labels. In this example, all yellow cells would be highlighted. If Labels are listed in the first row, as they are here, then check the box "Labels in First Row."

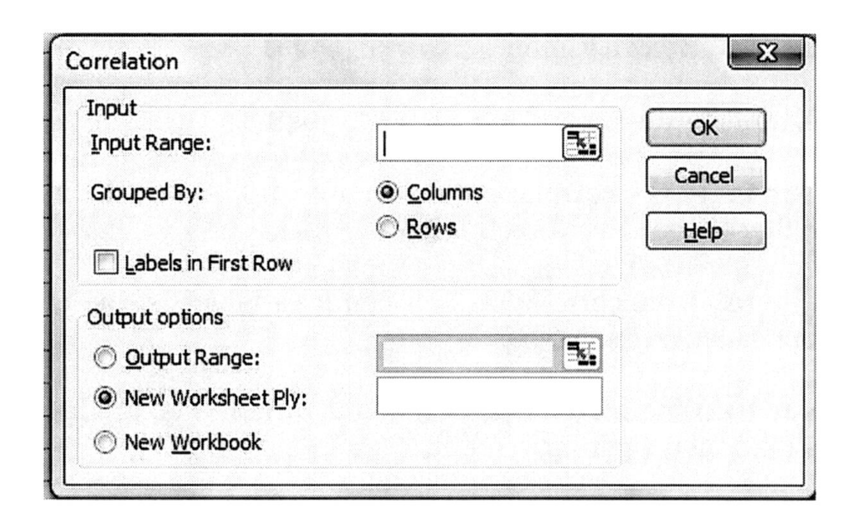

The output is a correlation matrix as shown below:

	x	y	z	a
x	1			
y	0.941884	1		
z	-0.50614	-0.39225	1	
a	0.545892	0.587241	-0.15648	1

We assume this data is sample data so the Correlation Coefficient is r, not **ρ** which is the Correlation Coefficient for data from a known population.

The correlation between x and y = $\mathbf{r}_{xy}$ = 0.941884

The correlation between x and z = $\mathbf{r}_{xz}$ = -0.50614

The correlation between x and a = $\mathbf{r}_{xa}$ = 0.545892

The correlation between y and z = $\mathbf{r}_{yz}$ = -0.39225

The correlation between y and a = $\mathbf{r_{ya}}$ = 0.587241

The correlation between z and a = $\mathbf{r_{za}}$ = -0.15648

The closer to +1 the correlation between 2 variables is, the more they move together in the same direction.

The closer to -1 the correlation between 2 variables is, the more they move in opposite directions.

The closer to 0 the correlation between 2 variables is, the less related and more random is their movement.

Covariance Analysis

The covariance also describes how linear a relationship is between two variables. The main difference between covariance and correlation is the range of values that each can assume. The Correlation between two variables can assume values only between -1 and +1. The Covariance between two variables can assume a value outside of this range. The more positive a covariance is, the more closely the variables move in the same direction. Conversely, the more negative a covariance is, the more the variables move in opposite directions.

Calculation of Covariance

Below is the formula for calculating the Covariance of variables. Excel does such a great job in calculating correlation and covariance that it is not necessary to memorize the formulas of covariance and correlation, but here they are, along with examples worked out in Excel:

Covariance of variables x and y from a known population = σ_{xy}

$\sigma_{xy} = 1/n * \Sigma (x_i - \mu_x) * (y_i - \mu_y)$ as i goes from 1 to n
μ_x and μ_y represent population means and s_{xy} represents a covariance from a population.

Covariance of variables x and y randomly sampled from an unknown population = s_{xy}
(This is the normal situation)

$s_{xy} = 1/ (n - 1) * \Sigma (x_i - x_{avg}) * (y_i - y_{avg})$ as i goes from 1 to n

Excel Functions Used When Calculating Covariance

COVAR (Highlighted block of cells of 2 variables)

= s_{xy} = Sample Correlation between two variables x and y

The Excel Statistical function **COVAR** calculates the covariance between 2 variables. The only inputs in the **COVAR** formula are specifying the locations of the blocks of cells for each variable. This is done by highlighting each block of cells after the function is inserted. **COVAR** is one of the statistical functions. **COVAR** can be used to calculate the covariance between only two variables. The Excel Data Analysis tool Covariance, which is discussed after **COVAR**, can calculate covariance between more than two variables simultaneously.

Here is an example of **COVAR** in use:

Problem 3: Calculating Covariance Between Two Variables

Problem: Calculate the covariance between variables x and y based upon the 6 pairs of x-y data given below:

x	y
1	2
3	6
6	7
8	9
5	6
4	5

The Covariance of variables x and y is s_{xy}:

s_{xy} = **COVAR (darker highlighted block, lighter highlighted block)** = 4.416667

Covariance (Highlighted block of multiple variables)

If you wish to calculate the covariance between more than 2 variables, you would use the Covariance data analysis tool. This can be found at:

Tools / Data Analysis / Covariance (in Excel 2010 -> Data tab)

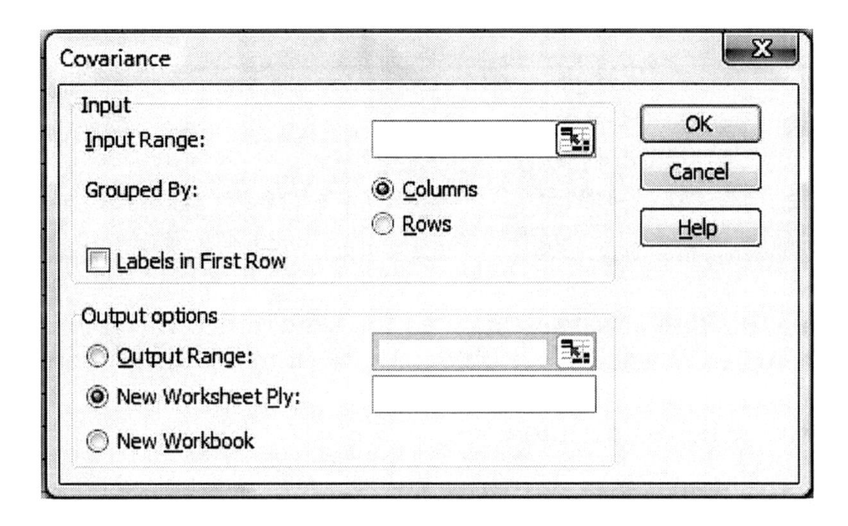

Problem 4: Calculating Covariance Between Multiple Variables

Problem: Determine the covariance between all of the variables below:

x	y	z	a
1	2	10	24
3	6	9	45
6	7	8	56
8	9	7	46
5	6	6	67
4	5	5	23

The only input for the Covariance function is to highlight all of the sample data and the row or column containing labels. In this example, all yellow cells would be highlighted. If Labels are listed in the first row, as they are here, then check the box "Labels in First Row."

The output is a covariance matrix as shown below:

	x	y	z	a
x	4.916667			
y	4.416667	4.472222		
z	-1.91667	-1.41667	2.916667	
a	19.25	19.75	-4.25	252.9167

We assume this data is sample data so the Covariance variable is $\mathbf{s_{xy}}$, not $\boldsymbol{\sigma_{xy}}$ which is the Covariance variable for data from a known population.

The covariance between x and y = $\mathbf{s_{xy}}$ = 4.416667

The covariance between x and z = $\mathbf{s_{xz}}$ = -1.91667

The covariance between x and a = $\mathbf{s_{xa}}$ = 19.25

The covariance between y and z = $\mathbf{s_{yz}}$ = -1.41667

The covariance between y and a = $\mathbf{s_{ya}}$ = 19.75

The covariance between z and a = $\mathbf{s_{za}}$ = -4.25

The more positive the covariance between 2 variables is, the more they move together in the same direction.

The more negative the covariance between 2 variables is, the more they move in opposite directions.

Hand Calculation of Combination and Permutation Problems

Go To
http://excelmasterseries.com/Excel_Statistical_Master/Correlation-Covariance.php

To View How To Solve Correlation and Covariance Problems By Hand (No Excel)

(Is Your Internet Connection Turned On ?)

You'll Quickly See Why You Always Want To Use Excel To Solve Statistical Problems !

Chapter 4 - The Normal Distribution

Basic Description of the Normal Distribution

The Normal distribution is one of the fundamental building blocks of statistics. The Normal distribution is also called the Gaussian distribution within the scientific community. Many measurements and physical phenomena can be approximated with the Normal distribution. The Normal Distribution can even be used to evaluate populations whose underlying distribution is unknown or even known not to be Normal. Statistics' most important theorem, the Central Limit Theorem, states the means of multi-point, representative samples will be normally distributed, regardless of the distribution of the underlying population. The Normal distribution is the most useful of all distributions.

Mapping the Normal Curve

The shape of the Normal distribution resembles a bell so it is sometimes called the "bell curve." The Normal curve is symmetric about its mean in the middle, and its tails on either side extend to infinity. The Normal distribution is a continuous function, not a discrete function, This means that any value can be graphed somewhere on a Normal curve. Discrete functions only map specific values, such as whole numbers.

Any Normal distribution can be completely mapped if only the following two parameters are known:

1) Mean - **μ** - the Greek symbol "mu"

2) Standard deviation - **σ** - the Greek symbol "sigma"

If the mean and standard deviation of a Normal distribution are known, then every point of the Normal curve can be mapped.

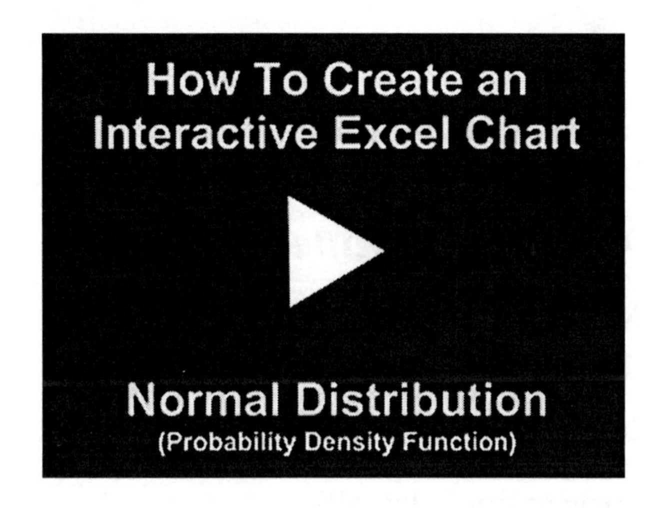

Instructional Video

Go to
http://www.youtube.com/watch?v=h4Zd7iGB4aI
to View a
Video From Excel Master Series
About How To Create
a User-Interactive Graph of the
Normal Distribution's
Probability Density Function
in Excel

(Is Your Internet Connection and Sound Turned On?)

The Standardized Normal Curve

A **Standardized Normal Curve** is a Normal curve that has **mean = 0** and **standard deviation = 1**

The "68 - 95 - 99.7%" Rule

The "68 - 95 - 99.7%" Rule for the Normal distribution states that:

68% of all observations lie within 1 standard deviation of the mean, within the range of μ +/- σ

95% of all observations lie within 2 standard deviations of the mean, within the range of μ +/- 2σ

99.7% of all observations lie within 3 standard deviations of the mean, within the range of μ +/- 3σ

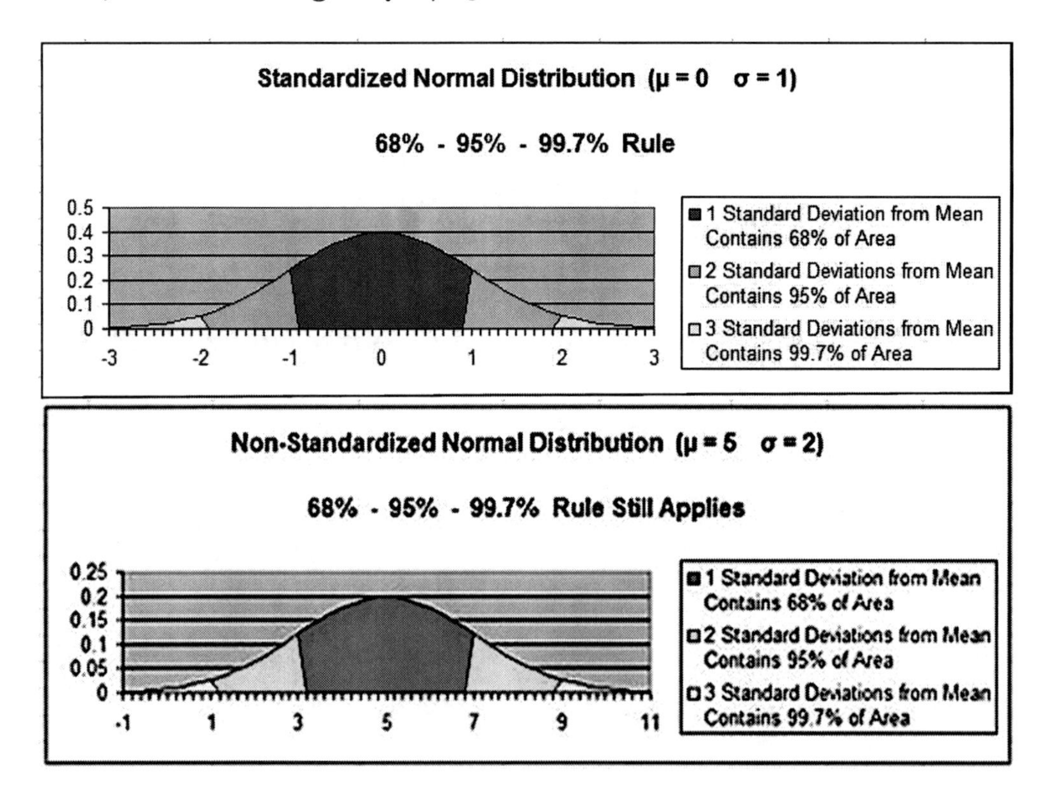

"Six Sigma Quality" in the Corporate World

As a little aside note - when companies state that they are trying to reach the "six sigma" level, they mean that they are trying to make their processes exact enough that no errors will occur inside of 4.5 standard deviations of output measurements. Processes that operate with "six sigma quality" are expected to produce defect levels below 3.4 defects per million production runs. 1.5 sigmas are customarily subtracted from the original six to take into account long-term variation. After a subtraction of 1.5 sigmas from the original 6, only 4.5 sigmas from the mean are actually used to obtain the "6 sigma" quality standard.

The 4 Most Important Excel Normal Functions

NORMDIST

NORMDIST(x, mean, standard dev, TRUE) = % area to the left of x

This function calculates the percentage of Normal curve area to the left of point x. Other inputs include the Normal curve's mean and standard deviation. TRUE must be stated as well. This type of function is a Cumulative Distribution function because it calculates the cumulative area under the Normal curve from -∞ in the left tail to point x. Stating TRUE as an input specifies that this is a Cumulative Distribution function.

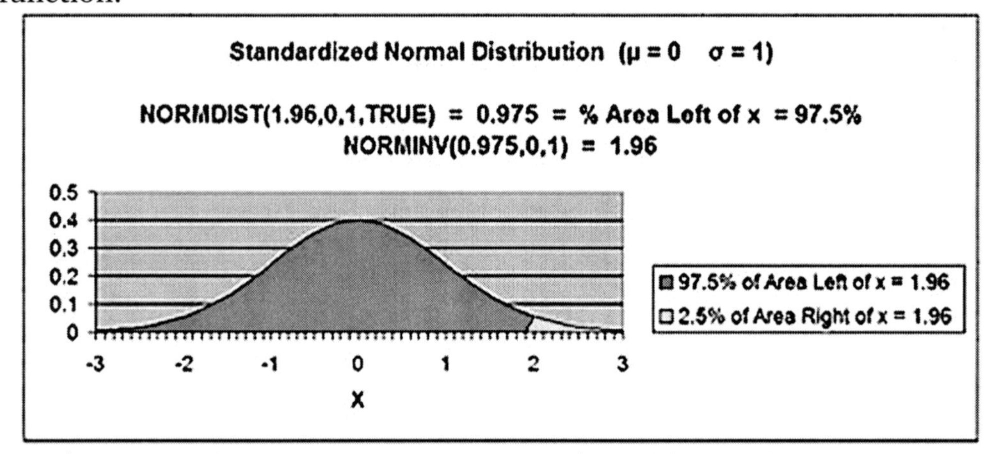

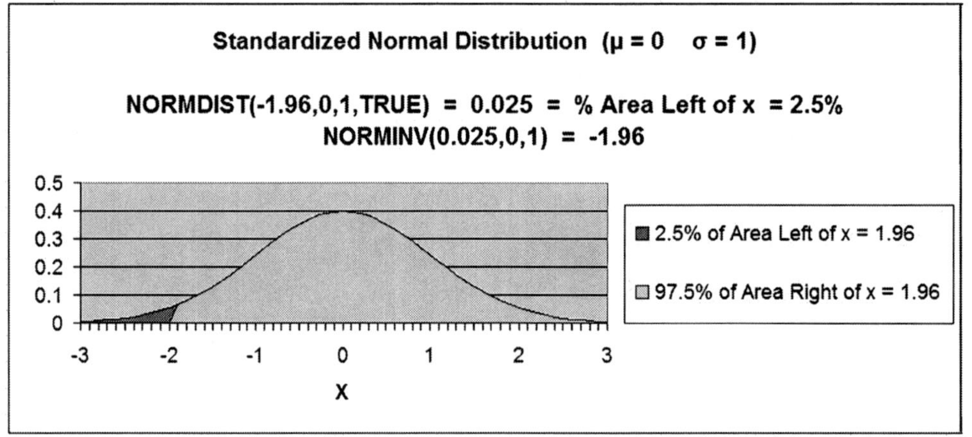

NORMSDIST

NORMSDIST(x) = % area to the left of x
This function calculates the percentage of Normal curve area to the left of point x standard deviations from the mean. This function assumes that the Normal curve is standardized, that is, has a mean = 0 and standard deviation = 1.

NORMSDIST(1.96) = 0.975
97.5% of the total area under the Normal curve lies left of a point 1.96 standard deviations to the right of the mean.

NORMSDIST(0) = 0.5
50% of the total area under the Normal curve lies left of the mean.

NORMINV

NORMINV(% of area to the left of x, mean, standard deviation) = x
This function tells where point x is given the inputs of percentage of area to left of x, the mean, and standard deviation of this Normal curve.
NORMINV(0.975,0,1) = 1.96

NORMSINV

NORMSINV(% of area to the left of x) = **x**
This function states that the given percentage of area under the Normal curve (the user input of this function) will be to the left of a point that is **x** standard deviations from the mean. **x** equals the number of standard deviations because the function NORM**S**INV assumes that the Normal distribution is standardized with mean = 0 and standard deviation = 1.

If this example, 97.5% of the area under the Normal curve Is to the left of the point that is 1.96 standard deviations from the mean:

NORM**S**INV(0.975) = 1.96

Problem 1: Using the Normal Distribution to Determine the Probability of Daily Sales Below a Certain Point

Problem: A store has normally distributed daily sales. The average daily sales = $2,000 and the daily sales standard deviation = $500, What is the probability that the sales during one day will fall below $1,000?

Since we are solving for a probability (area under the Normal curve) for a Normal curve that is not standardized, use NORMDIST. instead of NORMSDIST, which is used for a standardized Normal curve.

(Standardized Normal curve has mean = 0 and standard deviation = 1)

NORMDIST (1000,2000,500,TRUE) = 0.0228

Only 2.28% of the total area under this Normal curve falls to the left of x = 1,000 if the mean = 1,000 and the standard deviation = 500

Therefore the probability is only 2.28% that daily sales will fall below 1,000. This is shown on the following diagram on the next page.

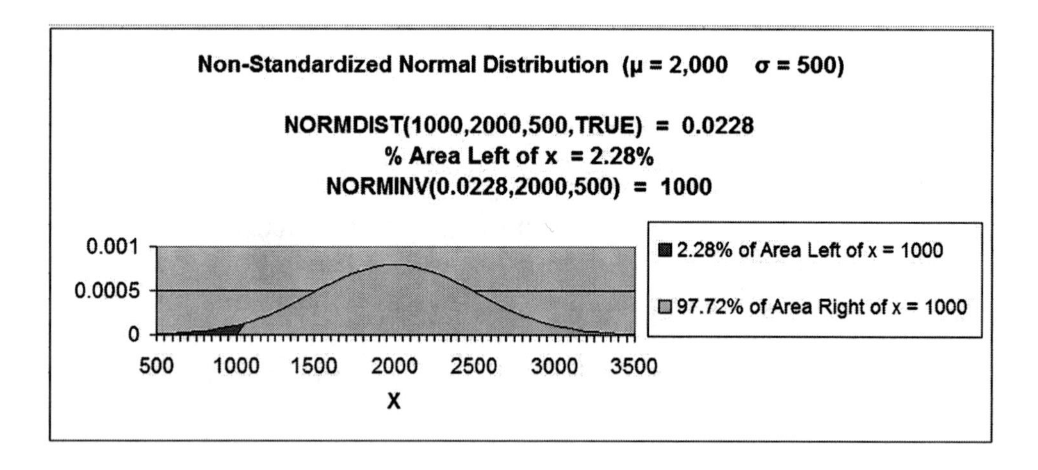

Problem 2: Using the Normal Distribution to Determine the Probability that Fuel Consumption is within a Certain Range

Problem: A brand of car has a mean fuel consumption of 27 mpg with a standard deviation of 5 mpg. What percentage of the cars can be expected to have a fuel consumption of between 25 mpg AND 30 mpg? Fuel consumption is normally distributed for this population.

Population Mean = μ = "mu" = 27 mpg

Population Standard Deviation = σ = "sigma" = 5 mpg

25 mpg ≤ x ≤ 30 mpg

Probability that 25 mpg ≤ x ≤ 30 mpg = ?

mpg is Normally distributed

Normal curve is not standardized (μ ≠ 0, σ ≠ 1)

Population Mean = **μ** = "mu" = 27 mpg

Population Standard Deviation = **σ** = "sigma" = 5 mpg

25 mpg ≤ x ≤ 30 mpg

Probability that 25 mpg ≤ x ≤ 30 mpg = ?

mpg is Normally distributed

Normal curve is not standardized (**μ** ≠ 0, **σ** ≠ 1)

Since we are solving for a probability (area under the Normal curve) for a Normal curve that is not standardized, use NORMDIST, not NORM**S**DIST.

Probability of a car having fuel efficiency between 25 mpg **AND** 30 mpg

(The AND statement requires that the answer contain the Union (the elements common to both sets). This will mean subtracting the smaller set from the bigger set to obtain only the probabilities common to both sets.)
= Probability of a car having fuel efficiency less than 30% - Probability of a car having fuel efficiency less than 25% =

= NORMDIST(30,27,5,TRUE) - NORMDIST(25,27,5,TRUE) =

= 0.72547 - 0.344578 = 0.381169

38.1% of all cars can be expected to have fuel efficiency between 25 mpg and 30 mpg.

This problem can be more clearly understand by observing the following diagrams on the next page:

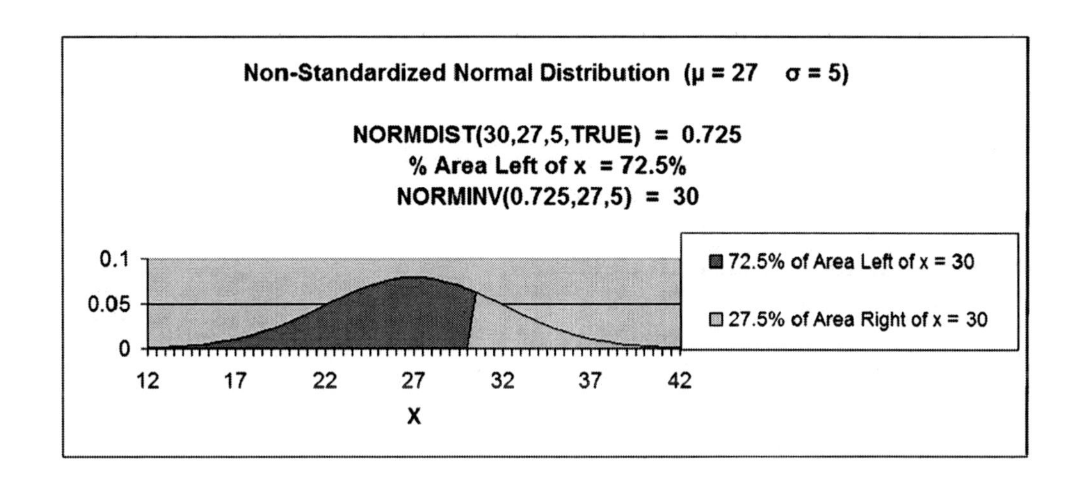

Non-Standardized Normal Distribution (μ = 27 σ = 5)
NORMDIST(30,27,5,TRUE) = 0.725
% Area Left of x = 72.5%
NORMINV(0.725,27,5) = 30
72.5% of Area Left of x = 30
27.5% of Area Right of x = 30
0.1
0.05
0
12 17 22 27 32 37 42
X

-

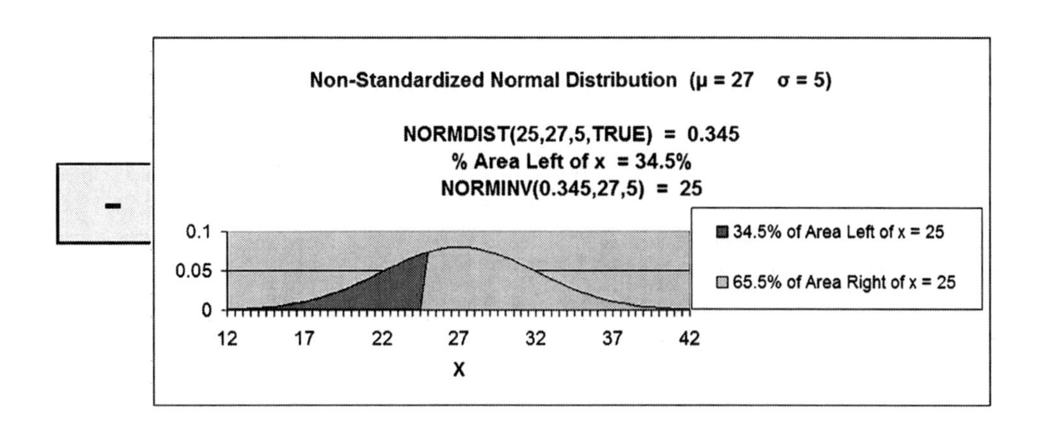

Non-Standardized Normal Distribution (μ = 27 σ = 5)
NORMDIST(25,27,5,TRUE) = 0.345
% Area Left of x = 34.5%
NORMINV(0.345,27,5) = 25
34.5% of Area Left of x = 25
65.5% of Area Right of x = 25
0.1
0.05
0
12 17 22 27 32 37 42
X

=

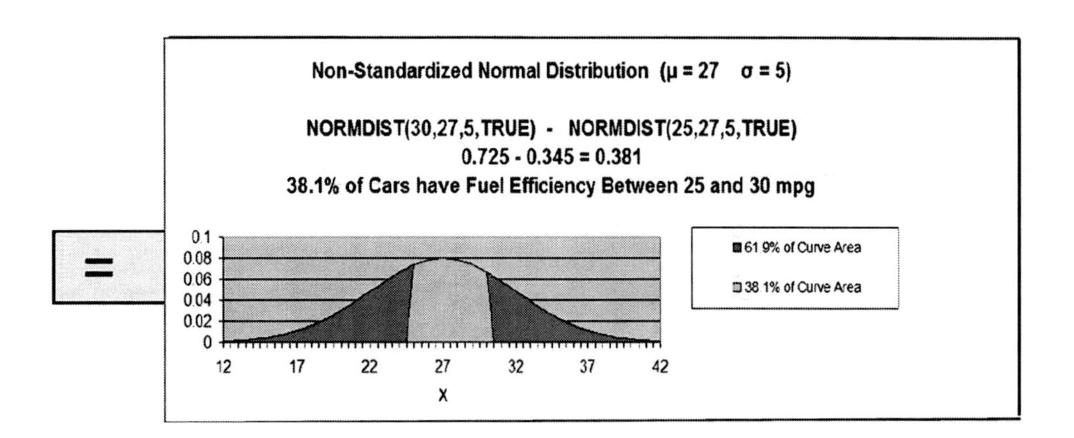

Non-Standardized Normal Distribution (μ = 27 σ = 5)
NORMDIST(30,27,5,TRUE) - NORMDIST(25,27,5,TRUE)
0.725 - 0.345 = 0.381
38.1% of Cars have Fuel Efficiency Between 25 and 30 mpg
61.9% of Curve Area
38.1% of Curve Area
0.1
0.08
0.06
0.04
0.02
0
12 17 22 27 32 37 42
X

Problem 3: Using the Normal Distribution to Determine the Upper Limit of Delivery Time

Problem: A company's package delivery time is normally distributed with a mean of 10 hours and a standard deviation of 3 hours. What delivery time will be beaten by only 2.5% of all deliveries?

If solving for x given a probability of x for a Normal curve that is not standardized, use NORMINV, not NORM**S**INV.

Note that NORMINV calculates the area to the left of x, so the problem parameters must be written calculating a probability less than x.

NORMINV(**0.025**,10,3) = 4.12

Only 2.5% of all package delivery times will be quicker than 4.12 hours.

In this chart, x = Delivery time measured in hours.

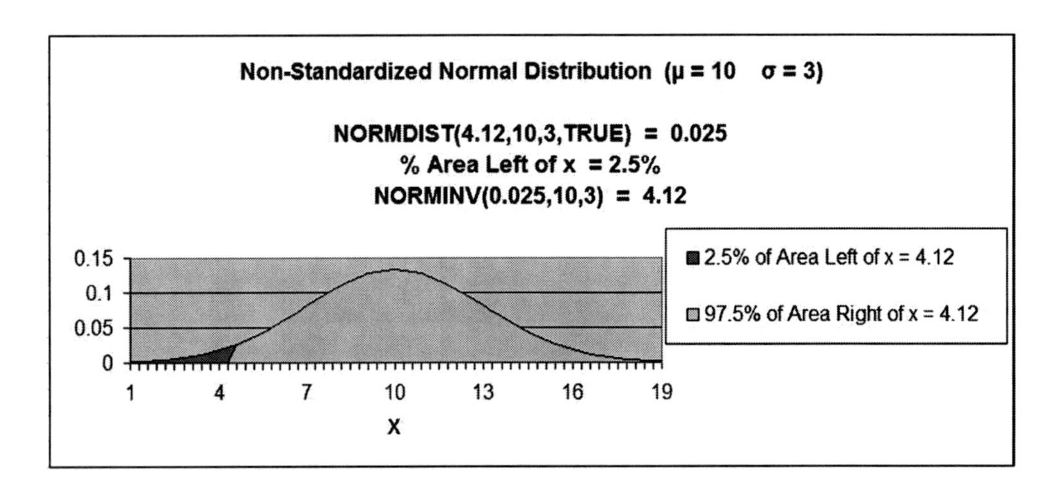

Problem 4: Using the Normal Distribution to Determine the Lower Limit of Tire Life

Problem: A tire company makes a tire with a normally distributed tread life that has a mean of 39,000 miles and standard deviation of 5,300 miles. What tread life would be exceeded by 98% of all tires?

(If solving for x given the probability of x for a Normal curve that is not standardized, use NORMINV, not NORM**S**INV)

Note that **NORMINV calculates the area to the left of x**, so the problem parameters must be written calculating a **probability less than x**. The problem is originally worded asking for the probability of greater than x.

NORMINV(**0.02**,39000,5300) = 28,115

Only 2% of all tires will wear out before 28,115 miles.

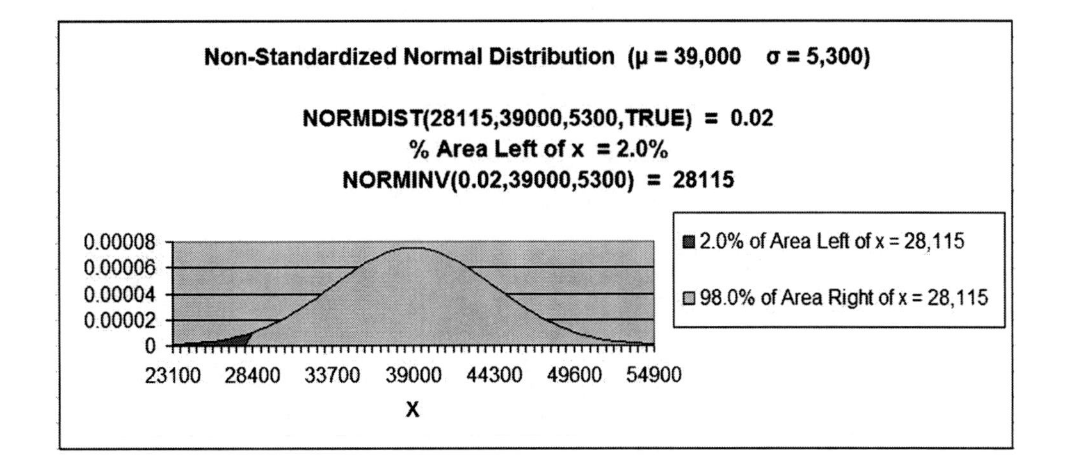

Problem 5: Using the Normal Distribution to Determine the Boundaries of a Range of Tire Life

Problem: A tire company makes a tire with a normally distributed tread life that has a mean of 39,000 miles and standard deviation of 5,300 miles. What would the range of tread life be that 95% of all tires would wear out in?

Population Mean = μ = "mu" = 39,000 miles

Population Standard Deviation = σ = "sigma" = = 5,300 miles

x_1 = ? x_2 = ?

<u>Left Boundary</u>
Probability that (tread life ≤ x_1) = 2% = **0.025**

And

<u>Right Boundary</u>
Probability that (tread life ≤ x_2) = 97.5% = **0.975**

Tread life is Normally distributed

Normal curve is not standardized (μ ≠ 0, σ ≠ 1)

If solving for x given a probability of x for a Normal curve that is not standardized, use NORMINV.

Note that **<u>NORMINV calculates the area to the left of x</u>**, so the problem parameters must be written calculating a **<u>probability less than x.</u>**

A Range is normally bounded on its upper and lower ends. We arbitrarily chose to have the 95% range bounded with 2.5% of tires outside on the upper and 2.5% of tires outside on the lower end. We could have chosen any combination that would have 5% of tires outside, such as 1% on top and 4% on the bottom.

Calculation of the left boundary:

NORMINV(**0.025**,39000,5300) = 28,612

Only **2.5%** of all tires will wear out before **28,115** miles.

Calculation of the right boundary:

NORMINV(**0.975**,39000,5300) = **49,388**

Only 2.5% of all tires will wear out after 49,388 miles. (**97.5%** of all tires will wear out before 48,388 miles)

So, 95% of tires will wear out in the range of **28,612** miles to **49,388** miles.

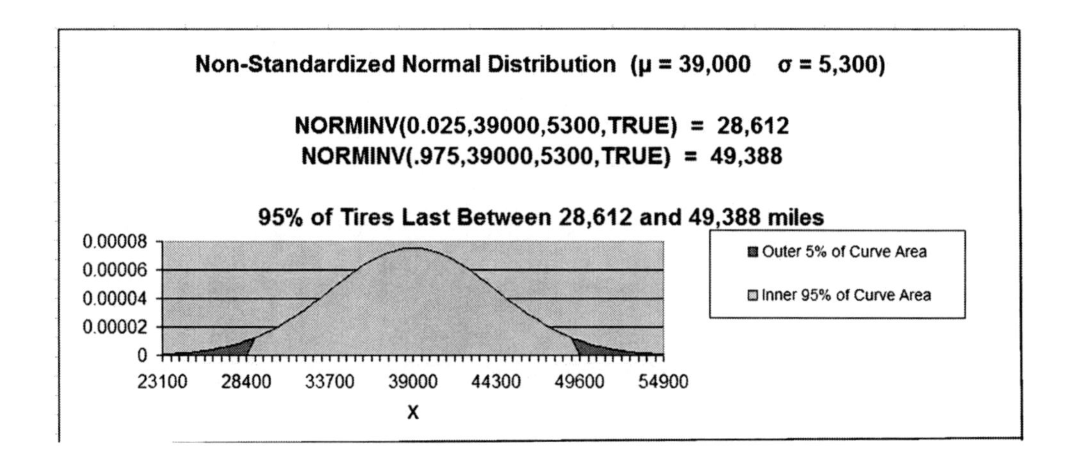

Problem 6: Using the Normal Distribution to Determine the Probability of a Pumpkin's Weight Being in 1 of 2 Ranges

Problem: Assume that X, the weight of pumpkins on a farm, is normally distributed with a mean of 20 pounds and a standard deviation of 1 pound. A pumpkin is selected at random, and its weight is noted. What is the probability that the weight will be either more than 21.50 pounds OR less than 18.5 pounds?

Population Mean = μ = "mu" = 20 lbs

Population Standard Deviation = σ = "sigma" = = 1 lb

18.5 lbs ≥ x OR x ≥ 21.5 lbs mpg

Probability that 18.5 lbs ≥ x OR x ≥ 21.5 lbs mpg = ?

weight is Normally distributed

Normal curve is not standardized (μ ≠ 0, σ ≠ 1)

Since we are solving for a probability (area under the Normal curve) for a Normal curve that is not standardized, use NORMDIST, not NORM**S**DIST.

Probability of a pumpkin weighing less than 18.5 lbs **OR** more than 21.5 lbs =

The OR statement requires that the answer contain all elements of both sets). This will mean adding both sets together and then subtracting any repeating elements. Both sets are mutually exclusive so no elements are repeating in both sets. We therefore just add both sets together.

The probability of a pumpkin weighing **less than** 18.5 lbs = NORMDIST(18.5,20,1,TRUE) = **0.066807** = 6.67%

OR

The probability of a pumpkin weighing **more than** 21.5 lbs = **1 -** NORMDIST(21.5,20,1,TRUE)

=1 - 0.933193 = **0.066807**

= 6.67%

= Probability of a pumpkin weighing **less than** 18.5 lbs + Probability of a pumpkin weighing **more than** 21.5 lbs. =

= NORMDIST(18.5,20,1,TRUE) + [**1** - NORMDIST(21.5,20,1,TRUE)] =

= **0.066807** + **0.066807** = 0.133614 = 13.36%

So the probability that the pumpkin weighs less than 18.5 lbs or more than 21.5 lbs = 13.36%

Normality Tests - When To Use Them

Normality is a requirement in most parametric tests done in marketing. These would include statistical tests that involve the normal distribution, the t distribution, the chi-square distribution, and the F distribution. In fact, any test that is not a nonparametric test usually has some requirement of normality.

Parametric statistical tests that require normality include the t test, the z test, ANOVA, correlation, covariance, regression, the chi-square test of independence, and the chi-square test of population variance, and F tests. Each requires normality as follows:

Z tests - These explicitly require normally-distributed variables because Z scores are drawn from the normal distribution.

t - tests - Each of the two populations being compared must be normally distributed.

ANOVA - Each of the two or more populations from which samples are drawn must be normally distributed.

Regression - The residuals must be normally distributed.

chi-square tests - Require samples drawn from normally-distributed populations.

F tests - Since the F distribution is the ratio of chi-squared variables divided by their individual degrees of freedom, and the chi-square distribution required normally-distributed variables, F tests also require normally-distributed variables.

The F distribution is the ratio of two independent chi-squared variables divided by their respective degrees of freedom, and since the chi-square distribution requires a normal distribution, the F distribution is also going to require a normal distribution.

Hand Calculation of Normal Distribution Problems

Go To
http://excelmasterseries.com/Excel_Statistical_Master/Normal-Distribution.php

To View How To Solve Normal Distribution Problems By Hand (No Excel)
(Is Your Internet Connection Turned On ?)

You'll Quickly See Why You Always Want To Use Excel To Solve Statistical Problems !

Chapter 5 - t Distribution & Small Samples

Basic Description of the t Distribution

The t Distribution is often applied to small samples (**n**<30) to estimate a Confidence Interval of the mean for a much larger population. It is also used to perform a Hypothesis test called a t-test which determines whether the means of two groups are statistically different from each other.

The t Distribution curve looks very much like the bell-shaped Normal Distribution curve, except it is lower, flatter, and wider. In the real world, data often has heavier tails than the Normal Distribution describes. This is often caused by outliers. If it difficult to remove or downweigh the outliers, then the t Distribution is often used to replace the Normal Distribution.

The t Distribution is also called the Student's t Distribution.

Degrees of Freedom

The shape of a t Distribution depends upon sample size, n. An important parameter related to sample size is the Degrees of Freedom, or **v** ("nu") --> (also sometimes written as df)

Degrees of Freedom = **v** = **n** - 1

The smaller the Degrees of Freedom, the lower and flatter is the t Distribution curve. The lowest, flattest t Distribution occurs at **v** = 1 (sample size, **n**, equals 2). As the Degrees of Freedom increase, the t Distribution becomes higher and more peaked to approach the shape of the Normal Distribution curve. At approximately **n** = 30, the t Distribution begins to very closely resemble the Normal Distribution.

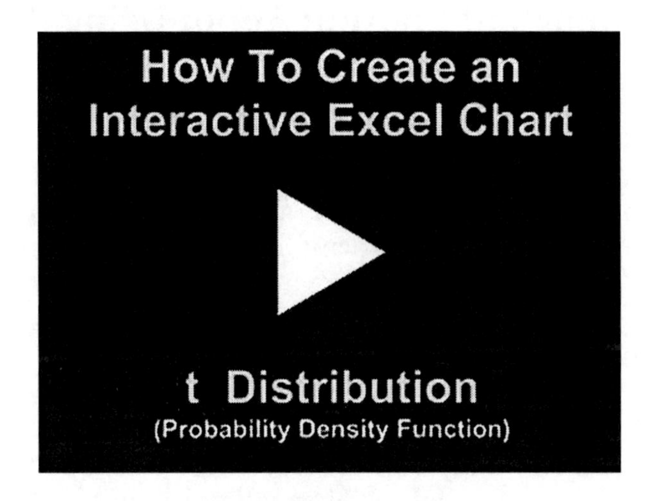

Instructional Video

Go to
http://www.youtube.com/watch?v=fqUo8-ykNbw
to View a
Video From Excel Master Series
About How To Create
a User-Interactive Graph of the
t-Distribution's
Probability Density Function
in Excel

(Is Your Internet Connection and Sound Turned On?)

One Very Important Caution About Using the t Distribution

Applying the t Distribution to small samples should only be done if it can be **proven** that the underlying population is Normally distributed. This is not usually the case. The derivation of the t Distribution is based upon the sample being drawn from a Normally distributed population. If the t Distribution is applied to small sample data to estimate the Confidence Interval of the mean of an underlying population that is not Normally distributed, the result can be **totally** wrong.

The statistical tests described in this manually require that the data being examined be Normally distributed. These tests should not be used unless it is known that the data is Normally distributed. Nonparametric test, which are not discussed in this manual, should be used in the event that it is not known if the data is Normally distributed.

Well-known tests of normality include the Shapiro-Wilk test and the Kolmogorov-Smirnoff test. These are not discussed within this course. Another well-known normality test, the Chi-Square Goodness-of-Fit test is a variation of the Chi-Square Independence Test, which is discussed in detail in an entire chapter near the end of this manual.

A Very Simple Excel Normality Test

A very simple normality test that can be run in Excel is to simply create a histogram of the data. If the shape of the histogram resembles the Normal distribution, you have a good indication that your data is Normally distributed. If the histogram shape has no resemblance to the Normal distribution (and you have at least 50 samples), your data is probably not Normally distributed. The histogram in Excel is so quick and simple that it should be a first step in normality testing.

One Convenient Way Around The Normality Requirement

One way around the requirement of Normality is to perform statistical testing on the means of multi-point samples. If, for example, each sample consists of the same number of multiple data points collected randomly and representatively, then the means of those samples will be Normally distributed. Statistics' most basic theorem, the Central Limit Theorem, states this. You can actually verify this in Excel. Use Excel's random number generator to generate 1,000 random numbers between 0 and 1. Divide the 1,000 numbers into groups of five numbers each. You will have 200 groups. Take the mean (average) of each group. If you create a histogram of these means in Excel, the histogram will be shaped just like the Normal curve. It is best to use at least 1,000 random numbers for this experiment.

I highly recommend that readers try this experiment. You will have a much stronger grasp of statistics' most powerful theorem if you do. You will also understand how you can manipulate data from any population, regardless of underlying distribution, so that you can run all of the Normal distribution-based tests described in this manual. Very useful stuff. If you don't know how to perform nonparametric testing, this will be extremely useful and practical information for you.

Estimating Confidence Interval with the t Distribution

This course module will provide a basic explanation of the calculations of a population's Confidence Interval of the mean from small sample data using the t Distribution. A more detailed explanation of estimating a Confidence Interval can be found in the course module entitled "Confidence Intervals."

Levels of Confidence and Significance

Level of Significance, **α** ("alpha"), equals the maximum allowed percent of error. If the maximum allowed error is 5%, then **α** = 0.05.

Level of Confidence is selected by the user. A 95% Confidence Level is the most common. A 95% Confidence Level would correspond to a 95% Confidence Interval of the Mean. This would state that the actual population mean has a 95% probability of lying within the calculated interval. A 95% Confidence Level corresponds to a 5% Level of Significance, or **α** = 0.05. The Confidence Level therefore equals 1 - **α**.

Population Mean vs. Sample Mean

Population Mean = **μ** ("mu") (This is what we are trying to estimate)

Sample Mean = x_{avg}

Standard Deviation and Standard Error

Standard Deviation is a measure of statistical dispersion. It's formula is the following: SQRT ([SUM $(x - x_{avg})^2$] / N). Standard Deviation equals the square root of the Variance. There is no need to memorize the formula because you can plug in Excel's STDEV function discussed below.

Population Standard Deviation = **σ** ("sigma")

Sample Standard Deviation = s

Standard Error is an estimate of population Standard Deviation from data taken from a sample. If the population Standard Deviation, **σ**, is known, then the Sample Standard Error, s_{xavg}, can be calculated. If only the Sample Standard Deviation, **s**, is known, then Sample Standard Error, s_{xavg}, can be estimated by substituting Sample Standard Deviation, **s**, for Population Standard Deviation, **σ**, as follows:

Sample Standard Error = s_{xavg} = **σ** / SQRT(**n**) ≈ **s** / SQRT(**n**)

σ = Population standard deviation
s = Sample standard deviation
n = sample size

Region of Certainty vs. Region of Uncertainty

Region of Certainty is the area under the Normal curve that corresponds to the required Level of Confidence. If a 95% percent Level of Confidence is required, then the Region of Certainty will contain 95% of the area under the Normal curve. **The outer boundaries of the Region of Certainty will be the outer boundaries of the Confidence Interval.**

The Region of Certainty, and therefore the Confidence Interval, will be centered about the mean. Half of the Confidence Interval is on one side of the mean and half on the other side.

Region of Uncertainty is the area under the Normal curve that is outside of the Region of Certainty. Half of the Region of Uncertainty will exist in the right outer tail of the Normal curve and the other half in the left outer tail. This is similar to the concept of the "two-tailed test" that used in Hypothesis testing in further sections of this course. The concepts of one and two-tailed testing are not used when calculating Confidence Intervals. Just remember that the Region of Certainty, and therefore the Confidence Interval, are always centered about the mean on the Normal curve.

Relationship Between Region of Certainty, Uncertainty, and Alpha

The Region of Uncertainty corresponds to α ("alpha"). If $\alpha = 0.05$, then that Region of Uncertainty contains 5% of the area under the Normal curve. Half of that area (2.5%) is in each outer tail. The 95% area centered about the mean will be the Region of Certainty. The outer boundaries of this Region of Certainty will be the outer boundaries of the 95% Confidence Interval. The Level of Confidence is 95% and the Level of Significance, or maximum error allowed, is 5%.

Degrees of Freedom

v = df = Degrees of Freedom = **n** - 1

The number of Degrees of Freedom is the main parameter of the t Distribution.

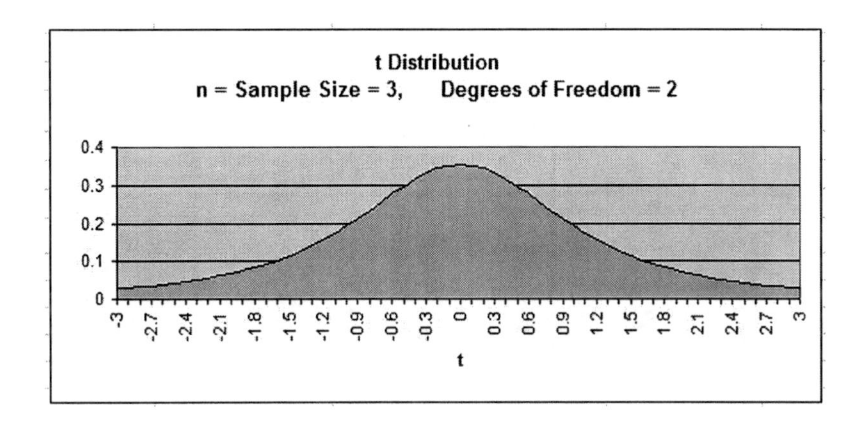

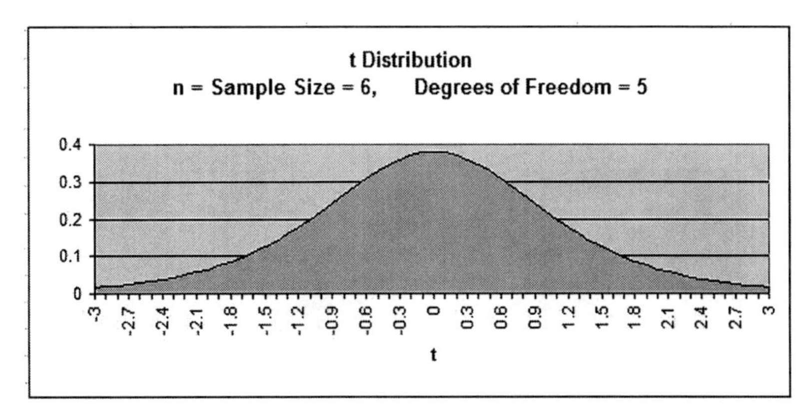

The lower the Degrees of Freedom, the fatter are the outer tails and the lower is the peak of the curve.

As the Degrees of Freedom increase, the tails get thinner and the peak of the curve gets higher. After the Degrees of Freedom exceed 30, the t-Distribution closely resembles the Normal curve, as is shown below:

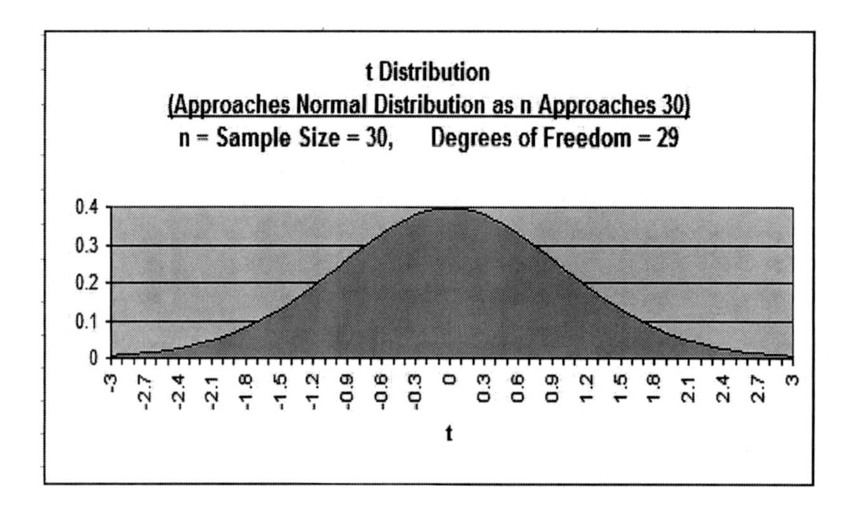

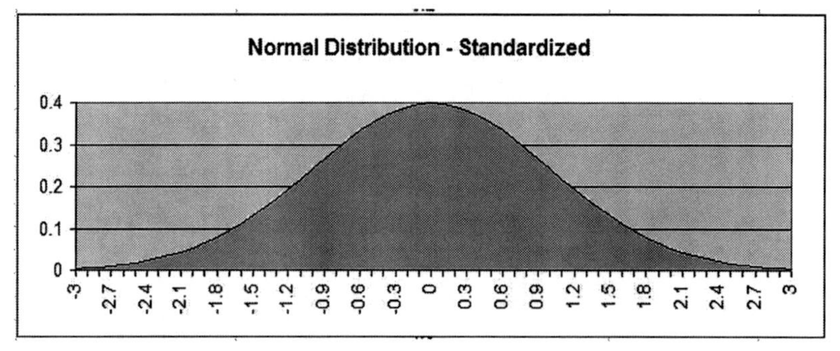

t Value

t Value is the number of Standard Errors from the mean to outer right boundary of the Region of Certainty (and therefore to the outer right boundary of the Confidence Interval). Standard Errors are used and not Standard Deviations because sample data is being used to calculate the Confidence Interval.

t Value is calculated by the following Excel statistical function:

t $Value_{(1-\alpha)}$ = TINV (**α, v**) - This will be discussed very shortly.

Excel Functions Used When Calculating Confidence Interval

COUNT

COUNT (Highlighted block of cells) = Sample size = n
----> Counts number of cells in highlighted block

STDEV

STDEV (Highlighted block of cells) = Standard deviation
----> Calculates Standard Deviation of all cells in highlighted block

AVERAGE

AVERAGE (Highlighted block of cells) = Mean
----> Calculates the mean of all cells in highlighted block

TINV

TINV (α, v) = t Value(1 - α)
= Number of Standard errors from mean to boundary of Confidence Interval.

For example:

Level of Confidence = 95% for a 95% Confidence Interval

Level of Significance = 5% (α = 0.05)

$1 - \alpha = 0.95 = 95\%$

If n = 6, then **v** = Degrees of Freedom = n - 1 = 5

t Value$_{95\%, v=5}$ = TINV (.05, 5) = 2.57

The outer right boundary of the 95% Confidence Interval, and the Region of Certainty, is 2.57 Standard Errors from the mean. The left boundary is the same distance from the mean because the Confidence Interval is centered about the mean.

Formula for Calculating Confidence Interval Boundaries from Sample Data Using the t Distribution

Confidence Interval Boundaries = Sample mean +/- t Value$_{(\alpha, v)}$ * Sample Standard Error

Confidence Interval Boundaries = $\mathbf{x_{avg}}$ +/- t Value$_{(\alpha, v)}$ * $\mathbf{s_{xavg}}$

Sample Mean = $\mathbf{x_{avg}}$ = AVERAGE (Highlighted block of cell containing samples)

t Value$_{(\alpha, v)}$ = TINV($\boldsymbol{\alpha}$,**v**)

Sample Standard Error = $\mathbf{s_{xavg}}$ = $\boldsymbol{\sigma}$ / SQRT(**n**) ≈ **s** / SQRT(**n**)

Sample size = **n** = COUNT (Highlighted block of cells containing samples)

Sample Standard Deviation = **s** = STDEV (Highlighted block of cells containing samples)

Problem: Calculate a Confidence Interval Based on Small Sample Data Using the t Distribution

Problem: Given the following set of 16 random test scores taken from a much larger population THAT IS NORMALLY DISTRIBUTED, calculate with 95% certainty an interval in which the population mean test score must fall. In other words, calculate the 95% Confidence Interval for the population test score mean using the t Distribution. Score samples are the following:

Test Scores:

220, 370, 500, 640, 220, 370, 500, 640, 220, 370, 500, 640, 220, 370, 500, 640

Level of Confidence = 95% = 1 - α

Level of Significance = α = 0.05

Sample Size = **n** = COUNT(Yellow Highlighted Block of Cells) = 16

Degrees of Freedom = **v** = **n** - 1 = 15

Sample Mean = x_{avg} = AVERAGE(Yellow Highlighted Block of Cells) = 432.5

Standard Deviation = **s** = STDEV(Yellow Highlighted Block of Cells) = 160.6

Sample Standard Error = s_{xavg} = σ / SQRT(**n**) ≈ **s** /SQRT(**n**)
= s_{xavg} = σ / SQRT(**n**) ≈ 160.6 / SQRT(16) = 40.2

t Value$_{(\alpha,v)}$ = TINV(α ,**v**) = TINV(0.05, 15) = 2.1314

Width of Half the Confidence Interval = t Value$_{(\alpha,v)}$ * s_{xavg}

= 2.1314 * 40.2 = 85.7

Confidence Interval Boundaries = $\mathbf{x_{avg}}$ +/- t Value$_{(\alpha,v)}$ * $\mathbf{s_{xavg}}$

= 432.5 +/- (2.1314)*(40.2) = 432.5 +/- 85.7

= 346.8 to 618.8

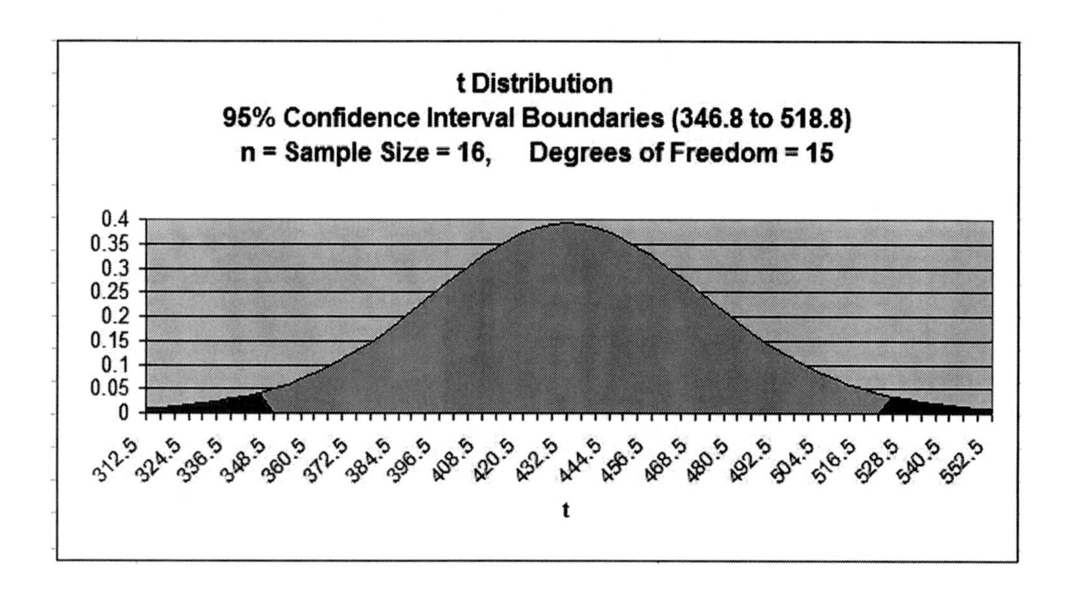

The t Test and Hypothesis Testing

The t Test is a Hypothesis test that is used to test whether the means of two groups are statistically different from each other. Quite often the t Test is applied to small sample data. As with other applications of the t Distribution to small sample data, the underlying population must be Normally distributed. This is often not the case.

The t Test will not be covered in this module because the two modules that cover Hypothesis testing provide significant detail about the topic.

The module of this course entitled "Excel Hypothesis Tools" provides examples of three different t Tests that are Data Analysis Tools of Excel. These are:

- t-Test: Paired Two Sample from Means
- t-Test: Two Sample Assuming Equal Variances
- t-Test: Two Sample Assuming Unequal Variances

Non-parametric tests can be used as alternatives to the t Test if large samples cannot be obtained and Normality of the underlying population cannot be proven. Some examples of applicable non-parametric tests would be the Mann-Whitney U test and the Wilcoxon test.

Hand Calculation of t Distribution Problems

Go To
http://excelmasterseries.com/Excel_Statistical_Master/t-Distribution.php

To View How To Solve t Distribution Problems By Hand (No Excel)

(Is Your Internet Connection Turned On ?)

You'll Quickly See Why You Always Want To Use Excel To Solve Statistical Problems !

Chapter 6 - Binomial Distribution

Instructional Video

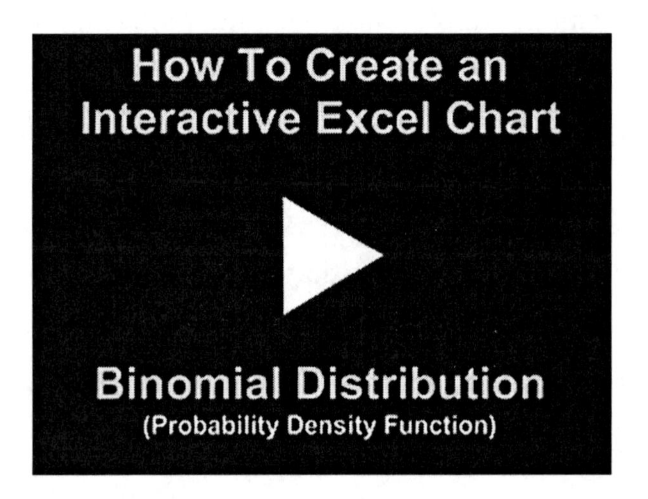

Instructional Video

Go to
http://www.youtube.com/watch?v=C8sUsVoxhZQ
to View a
Video From Excel Master Series
About How To Create
a User-Interactive Graph of the
Binomial Distribution's
Probability Density Function
in Excel

(Is Your Internet Connection and Sound Turned On?)

Basic Explanation of Binomial Distribution

The Binomial Distribution is one of the most valuable and commonly used statistical tools. The Binomial Distribution is used whenever a process has only two possible outcomes. For example, the Binomial Distribution would be used to determine the probability that 6 out of 10 people would vote Republican or the probability that 3 out of 20 people would have green eyes if the percentage of the overall population having green eyes or voting Republican were known.

The Binomial Distribution is applied to one sample of n trials taken from a much larger population. The Binomial Distribution will calculate the probability of the given number of successful outcomes in a given number of trials if the proportion of the overall population having that outcome is known.

As a general rule, the Binomial Distribution should only be applied to a sample if the population size, N, is at least 10 times larger than the sample size (n number of trials in the sample).

Bernoulli Trial

Bernoulli Trial is a single random experiment whose outcome can have only one of two possibilities: "success" or "failure." An example of this would be one flip of a coin. This is the same as one trial in a sample of n random trials whose outcomes are independent and have only two possibilities.

Bernoulli Process

Bernoulli Process is a sequence of Bernoulli Trials. An example of a Bernoulli Process would be 5 flips of a coin. A "Bernoulli Process" is just another name for a single sample of n random trials whose outcomes are independent and have only two possibilities. These are exactly the samples that the Binomial Distribution is applied to.

Bernoulli Distribution

Bernoulli Distribution is just another name for the Binomial Distribution applied to only 1 trial.

If you are trying to estimate a population proportion from the results of a sample proportion, see the module of this course entitled Confidence Intervals. Estimating a population proportion from a sample proportion is explained in the second half of that course module. In this module, we are doing almost the opposite of that. Here we are describing how to calculate the probability of a sample proportion if we know the population proportion.

Binomial Distribution Parameters

Random Variable

Random Variable = x
This is the variable that will have 1 of 2 possible outcomes.

Count of Successes per Trial

Count of Successes per n Trials = X

Population Proportion

Population Proportion = p = proportion of a population having a certain outcome or characteristic. p also equals the probability of a certain outcome on 1 trial.

q = 1 - p = proportion of a population not having that outcome or characteristic. q also equals the probability of not getting a certain outcome on 1 trial.

Sample Proportion

Sample Proportion = p_{avg} = proportion of a sample having a certain outcome or characteristic

$q_{avg} = 1 - p_{avg}$ = proportion of a sample not having that outcome or characteristic

Sample Size

Sample Size = n (This is the number of trials)

Expected Sample Occurrence Parameters

Expected Sample Occurrence Mean

Expected Sample Occurrence Mean = np

This is E(X) = the expected value of X (the count of successes in n trials)

Expected Sample Occurrence Variance

Expected Sample Occurrence Variance = npq

Expected Sample Occurrence Standard Deviation

Expected Sample Occurrence Standard Deviation = SQRT (npq)

Expected Sample Proportion Parameters

Expected Sample Proportion

Expected Sample Proportion = p

Expected Sample Proportion Variance

Expected Sample Proportion Variance = pq/n

Expected Sample Proportion Standard Deviation

Expected Sample Proportion Standard Deviation = SQRT (pq/n)

Probability Density Function
vs.
Cumulative Distribution Function

Probability Density Function

Probability Density Function = Pr (X = k)

The probability of exactly k successes in n trials. This is the probability that the count of successful outcomes, X, equals k for n trials. For example, the Probability Density Function would be used to calculate the probability of **exactly** 4 heads in 6 flips of a coin.

Pr (X = k) = n! / [k! * (n - k)!] * p^k * q^{n-k}

= BINOMDIST (k, p, n, FALSE)

Problem: Calculate the probability of exactly 4 heads in 6 flips of a coin.

Probability of success in each trial = p = 0.5

Number of trials = n = 6

Exact number of successes = k = 4

Pr (X = k) = **BINOMDIST (k, n, p, FALSE)**

= **BINOMDIST (4, 6, 0.5, FALSE)** = 0.234

Pr (X = k) = 23.4%

Cumulative Distribution Function

Cumulative Distribution Function = **Pr (X ≤ k)**

The probability of **UP TO** k successes in n trials. This is the probability that the count of successful outcomes, X, equals any number between k and 0 for n trials. This equals the sum of the probabilities that X equals each number from 0 to k for n trials. For example, the Cumulative Distribution Function would be used to calculate the probability of **up to** 4 heads in 6 flips of a coin;

Pr (X ≤ k) = $\text{Sum}_{\text{i from 0 to k}}$ Pr (X = i)

Pr (X ≤ k) = $\text{Sum}_{\text{i from 0 to k}}$ n! / [i! * (n - i)!] * p_i * q_{n-i}

4 = BINOMDIST (k, p, n, TRUE)

Problem 1: Probability of Getting a Certain Number of Successes for Binomial Variable Trials

Problem: Calculate the probability of up to 4 heads in 6 flips of a coin.

Probability of success in each trial = p = 0.5
Number of trials = n = 6
Maximum number of successes = k = 4

Pr (X ≤ k) = BINOMDIST (k, n, p, **TRUE**)
= BINOMDIST (4, 6, 0.5, TRUE) = 0.891

Pr (X ≤ k) = 89.1%

Problem 2: Probability of Getting a Certain Range of Successes for Binomial Variable Trials

Problem: What is the probability that between 10 and 25 products out of 100 require service if 15% of all products require service?

Pr (10 ≤ X ≤ 25) = Pr (X ≥ 10) **AND** Pr (X ≤ 25)

= Pr (X ≤ 25) **AND** Pr (X ≥ 10)
= PR (X ≤ 25) - Pr (X ≤ 9)

[Note that Pr (X ≥ 10) = 1 - Pr (X ≤ 9)]

= BINOMDIST (25, 100, 0.15, TRUE) - BINOMDIST (9, 100, 0.15, TRUE)
= 0.997 - 0.055 = 0.942 = 94.2%

There is a 94.2% probability that between 10 and 25 out of a random sample of 100 products will require service if 15% of all products require service.

Problem 3: Probability of Getting a Certain Range of Successes for Binomial Variable Trials

Problem: What is the probability of getting between 3 and 5 heads on 10 flips of a fair coin (p = 0.50) ?

Probability of getting between 3 and 5 heads = Pr(X=3) + Pr(X =4) + Pr(X =5)

Also equals [Pr(X=1) + Pr(X=2) + Pr(X=3) + Pr(X=4) + Pr(X=5)] - [Pr(X=1) + Pr(X=2)]

This equals [Cumulative probability of Pr(X ≤ 5)] - [Cumulative probability of Pr(X ≤ 2)]

= BINOMDIST (5,10,0.50, TRUE) - BINOMDIST (2, 10, 0.50, TRUE)

= 0.623 - 0.055 = 0.568

There is a 56.8% probability of getting between 3 and 5 heads on 10 flips of a fair coin.

Estimating the Binomial Distribution with the Normal and Poisson Distributions

The **Normal Distribution** can be used to approximate the Binomial Distribution if **n is large** and **p and q are not too close to 0**.

n	p	np	np(1-p)	X	Binomial (X, n, p)	Normal (X, μ, σ) = Normal (X, np, np(1-p))
1000	0.4	400	240	385	**0.0162**	**0.0161**
				400	**0.0257**	**0.0258**
				415	**0.0160**	**0.0161**

The **Poisson Distribution** can be used to approximate the Binomial Distribution if **n is large** and **p is small (less than 0.10).**

n	p	np	X	Binomial (X, n, p)	Poisson (X, λ) = Poisson (X, np)
100	0.03	3	1	**0.0162**	**0.0161**
			2	**0.0257**	**0.0258**
			3	**0.0160**	**0.0161**

The Binomial Distribution is a specific case of the more general Multinomial Distribution

Hand Calculation of Binomial Distribution Problems

Go To
http://excelmasterseries.com/Excel_Statistical_Master/Binomial-Distribution.php

To View How To Solve Binomial Distribution Problems By Hand (No Excel)

(Is Your Internet Connection Turned On ?)

You'll Quickly See Why You Always Want To Use Excel To Solve Statistical Problems !

Chapter 7 - Confidence Intervals

Basic Explanation of Confidence Intervals

Confidence Intervals are estimates of a population's average or proportion based upon sample data drawn from the population. A Confidence Interval is a range of values in which the mean is likely to fall with a specified level of confidence or certainty.

The Confidence Interval is an interval in which the true population mean or proportion probably lies based upon a much smaller random sample taken from that population.

Confidence Intervals for means are calculated differently than Confidence Intervals for proportions. The first half of this course module will discuss calculating a Confidence Interval for a population mean. The second half will cover calculating a Confidence Interval for a population proportion.

First we will briefly discuss the difference between sampling for mean and sampling for proportion:

Mean Sampling vs. Proportion Sampling

What determines whether a mean is being estimated or a proportion is being estimated is the number of possible outcomes of each sample taken.

Proportion samples have only two possible outcomes. For example, if you are comparing the proportion of Republicans in two different cities, each sample has only two possible values; the person sampled either is a Republican or is not.

Mean samples have multiple possible outcomes. For example, if you are comparing the mean age of people in two different cities, each sample can have numerous values; the person sampled could be anywhere from 1 to 110 years old.

Below is a description of how to calculate a Confidence Interval for a population's mean. Note that everything is almost the same as the calculation of the Confidence Interval for a population proportion, except sample standard error.

Confidence Interval of a Population Mean

The Confidence Interval of a Mean is an interval in which the true population mean probably lies based upon a much smaller random sample taken from that population. A 95% Confidence Interval of a Mean is the interval that has a 95% chance of containing the true population mean.

The width of a Confidence Interval is affected by the sample size. The larger the sample size, the more accurate and tighter is the estimate of the true population mean. The larger the sample size, the smaller will be the Confidence Interval. Samples taken must be random and also be representative of the population.

Calculate Confidence Intervals Using Large Samples (n>30)

Confidence Intervals are usually calculated and plotted on a Normal curve. If the sample size is less then 30, the population must be known to be Normally distributed. If small-sample data (n<30) is used to plot the Confidence Interval of the Mean for a population that is not Normally distributed, the result can be totally wrong.

Probably the most common major mistake in statistics is to apply Normal or t-distribution tests to small-sample data taken from a population of unknown distribution. Typically the actual distribution of a population is not known.

If the population's underlying distribution is not known (usually it is not), then only large samples (n>30) are valid for creating a Confidence Interval of the Mean. The most important theorem of statistics, the Central Limit Theorem, explains the reason for this.

The Central Limit Theorem

The Central Limit Theorem is statistics' most fundamental theorem. In a nutshell, it states the following: The means of random sample groups can be plotted on a Normal curve to estimate (create a confidence interval around) a population's mean no matter how the population is distributed, as long as sample size is large (n>30). No matter how the population is distributed, the sampling distribution of the mean approaches the Normal curve as sample size becomes large.

One Convenient Way Around The Normality Requirement

One way around the requirement of Normality is to perform statistical testing on the means of multi-point samples. If, for example, each sample consists of the same number of multiple data points collected randomly and representatively, then the means of those samples will be Normally distributed. Statistics' most basic theorem, the Central Limit Theorem, states this. You can actually verify this in Excel. Use Excel's random number generator to generate 1,000 random numbers between 0 and 1. Divide the 1,000 numbers into random groups of five numbers each. You will have 200 groups. Take the mean (average) of each group. If you create a histogram of these means in Excel, the histogram will be shaped just like the Normal curve. It is best to use at least 1,000 random numbers for this experiment.

I highly recommend that readers try this experiment. You will have a much stronger grasp of statistics' most powerful theorem if you do. You will also understand how you can manipulate data from any population, regardless of underlying distribution, so that you can run all of the Normal distribution-based tests described in this manual. Very useful stuff. If you don't know how to perform nonparametric testing, this will be extremely useful and practical information for you.

Levels of Confidence and Significance

Level of Significance, α ("alpha"), equals the maximum allowed percent of error. If the maximum allowed error is 5%, then **α** = 0.05.

Level of Confidence is the desired degree of certainty. A 95% Confidence Level is the most common. A 95% Confidence Level would correspond to a 95% Confidence Interval of the Mean. This would state that the actual population mean has a 95% probability of lying within the calculated interval. A 95% Confidence Level corresponds to a 5% Level of Significance, or α = 0.05. The Confidence Level therefore equals 1 - α.

Population Mean vs. Sample Mean

Population Mean = μ ("mu") (This is what we are trying to estimate)

Sample Mean = x_{avg}

Standard Deviation and Standard Error

Standard Deviation is a measure of statistical dispersion. It's formula is the following:

SQRT ([SUM $(x - x_{avg})^2$] / N).

There is no need to memorize the formula because you can plug in Excel's STDEV function discussed later. Standard Deviation equals the square root of the Variance.

Population Standard Deviation = σ ("sigma")

Sample Standard Deviation = s

Standard Error is an estimate of population Standard Deviation from data taken from a sample. If the population Standard Deviation, σ, is known, then the Sample Standard Error, s_{xavg}, can be calculated. If only the Sample Standard Deviation, s, is known, then Sample Standard Error, s_{xavg}, can be estimated by substituting Sample Standard Deviation, **s**, for Population Standard Deviation, **σ**, as follows:

Sample Standard Error = $s_{xavg} = \sigma$ / SQRT(**n**) ≈ **s** / SQRT(**n**)

σ = Population standard deviation
s = Sample standard deviation
n = sample size

Region of Certainty vs. Region of Uncertainty

Region of Certainty is the area under the Normal curve that corresponds to the required Level of Confidence. If a 95% percent Level of Confidence is required, then the Region of Certainty will contain 95% of the area under the Normal curve. **The outer boundaries of the Region of Certainty will be the outer boundaries of the Confidence Interval**.

The Region of Certainty, and therefore the Confidence Interval, will be centered about the mean. Half of the Confidence Interval is on one side of the mean and half on the other side.

Region of Uncertainty is the area under the Normal curve that is outside of the Region of Certainty. Half of the Region of Uncertainty will exist in the right outer tail of the Normal curve and the other half in the left outer tail. This is similar to the concept of the "two-tailed test" that is used in Hypothesis testing in further sections of this course. The concepts of one- and two-tailed testing are not used when calculating Confidence Intervals. Just remember that the Region of Certainty, and therefore the Confidence Interval, are always centered about the mean on the Normal curve.

Relationship Between Region of Certainty, Uncertainty, and Alpha

The Region of Uncertainty corresponds to α ("alpha"). If α = 0.05, then that Region of Uncertainty contains 5% of the area under the Normal curve. Half of that area (2.5%) is in each outer tail. The 95% area centered about the mean will be the Region of Certainty. The outer boundaries of this Region of Certainty will be the outer boundaries of the 95% Confidence Interval. The Level of Confidence is 95% and the Level of Significance, or maximum error allowed, is 5%.

Illustrating a two-tail test – Similar to what is used for calculating Confidence Interval

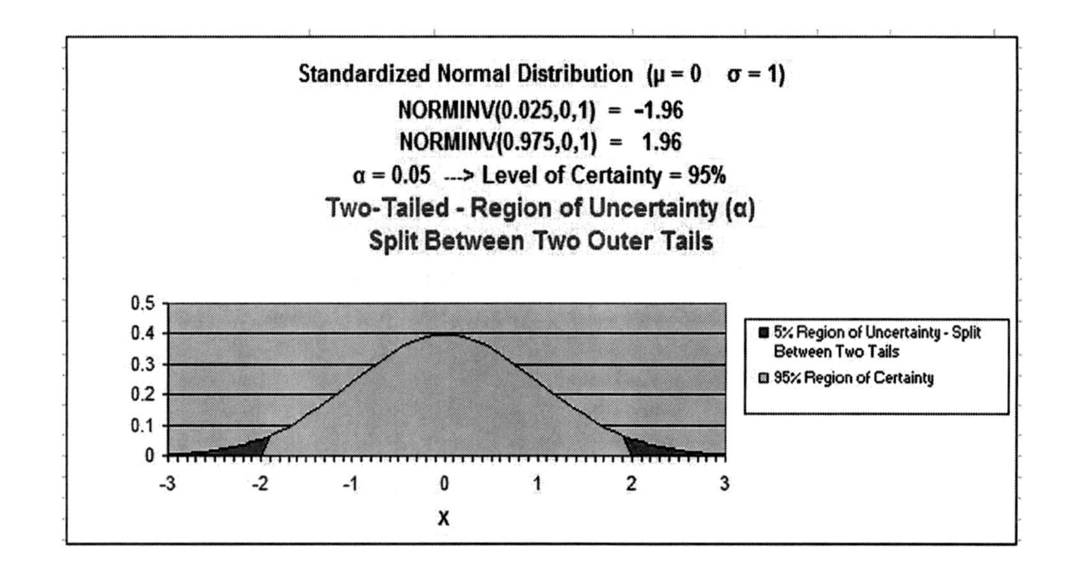

Illustrating a one-tailed test – Right Tail

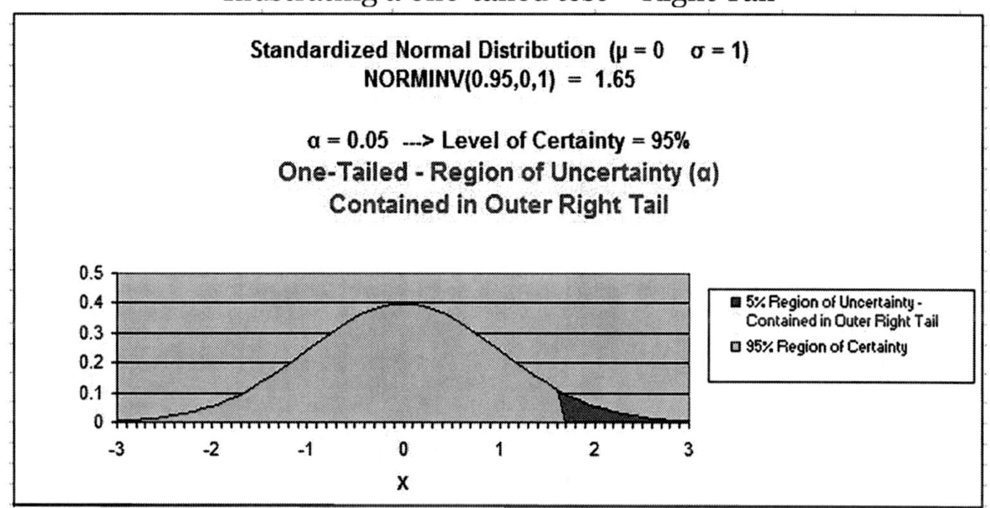

Illustrating a one-tailed test – Left Tail

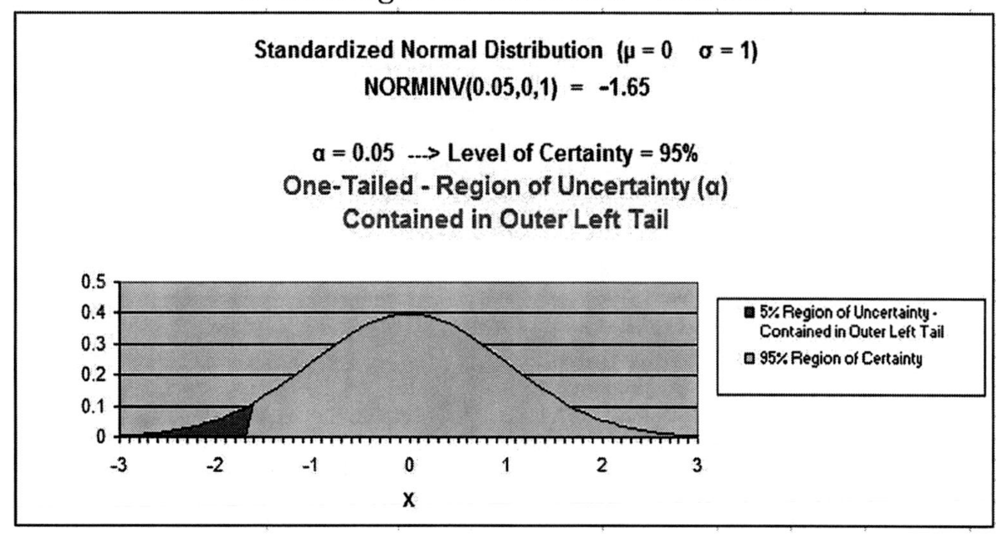

Z Score

Z Score is the number of Standard Errors from the mean to outer right boundary of the Region of Certainty (and therefore to the outer right boundary of the Confidence Interval). Standard Errors are used and not Standard Deviations because sample data is being used to calculate the Confidence Interval.

Z Score is calculated by the following Excel function:

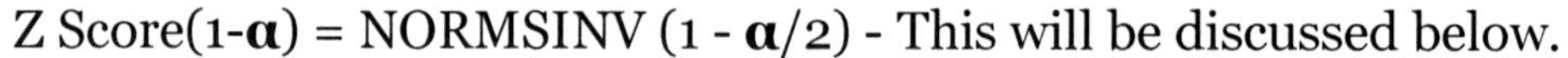

Z Score(1-α) = NORMSINV (1 - α/2) - This will be discussed below.

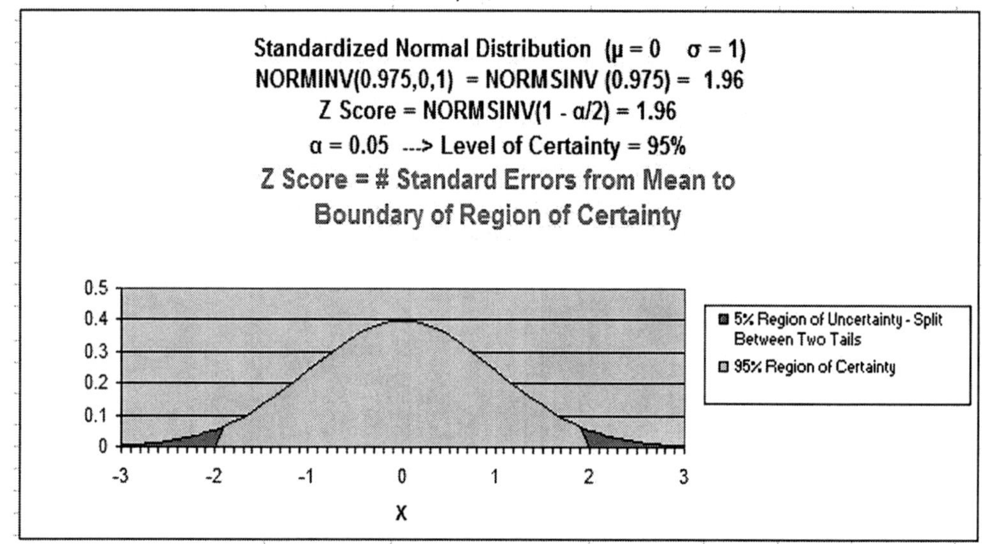

It is very important to note that on a Standardized Normal Curve, the Distance from the mean to boundary of the Region of Certainty equals the number of standard errors from the mean to boundary, which is the Z Score.

The above is only true for a Standardized Normal Curve. It is NOT true for a Non-Standardized Normal curve.

Excel Functions Used When Calculating Confidence Interval of Mean

COUNT

COUNT (Highlighted block of cells) = Sample size = n
----> Counts number of cells in highlighted block

STDEV

STDEV (Highlighted block of cells) = Standard deviation
----> Calculates Standard Deviation of all cells in highlighted block

AVERAGE

AVERAGE (Highlighted block of cells) = Mean
----> Calculates the mean of all cells in highlighted block

NORMSINV

NORMSINV (1 - α/2) = Z Score(1 - α)
= Number of Standard errors from mean to boundary of the Confidence Interval. Note that (1 - α/2) = the entire area in the Normal curve to the left of outer right boundary of the Region of Certainty, or Confidence Interval. This includes the entire Region of Certainty and the half of the Region of Uncertainty that exists in the left tail.

For example:

Level of Confidence = 95% for a 95% Confidence Interval

Level of Significance = 5% (α = 0.05)

1 - α = 0.95 = 95%

$Z\ Score_{95\%}$ = NORMSINV(1 – α/2) = NORMSINV (1 - .05/2) = NORMSINV(1 - 0.025)

$Z\ Score_{95\%}$ = NORMSINV (0.975) = 1.96

The outer right boundary of the 95% Confidence Interval, and the Region of Certainty, is 1.96 Standard Errors from the mean. The left boundary is the same distance from the mean because the Confidence Interval is centered about the mean.

CONFIDENCE

CONFIDENCE (**α**, **s**, **n**) = Width of half of the Confidence Interval

α = Level of Significance

s = Sample Standard Deviation - Note that this is not Standard Error.

s is calculated by applying STDEV to the sample values.

n = Sample size - Apply COUNT to sample values.

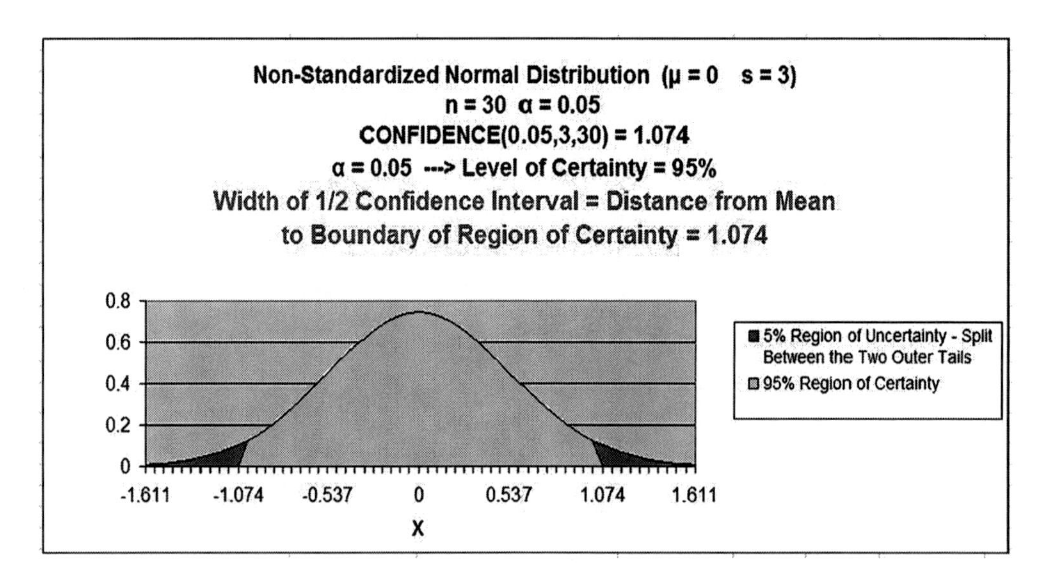

Formulas for Calculating Confidence Interval Boundaries from Sample Data

Confidence Interval Boundaries = Sample mean +/- Z Score$_{(1-\alpha)}$ * Sample Standard Error

Confidence Interval Boundaries = $\mathbf{x_{avg}}$ +/- Z Score$_{(1-\alpha)}$ * $\mathbf{s_{xavg}}$

Sample Mean = $\mathbf{x_{avg}}$ = AVERAGE (Highlighted block of cell containing samples)

Z Score$_{(1-\alpha)}$ = NORMSINV (1 - $\boldsymbol{\alpha}$/2)

Sample Standard Error = $\mathbf{s_{xavg}}$ = $\boldsymbol{\sigma}$ / SQRT(**n**) ≈ **s** / SQRT(**n**)

Sample size = **n** = COUNT (Highlighted block of cells containing samples)

Sample Standard Deviation = **s** = STDEV (Highlighted block of cells containing samples)

CONFIDENCE (α, s, n) = Width of half of the Confidence Interval

CONFIDENCE (α, s, n) = Z Score$_{(1-\alpha)}$ * $\mathbf{s_{xavg}}$

So:

Confidence Interval Boundaries = $\mathbf{x_{avg}}$ +/- Z Score$_{(1-\alpha)}$ * $\mathbf{s_{xavg}}$

Confidence Interval Boundaries = $\mathbf{x_{avg}}$ +/- CONFIDENCE (**α, s, n**)

Problem 1: Calculate a Confidence Interval from a Random Sample of Test Scores

Problem: Given the following set of 32 random test scores taken from a much larger population, calculate with 95% certainty an interval in which the population mean test score must fall. In other words, calculate the 95% Confidence Interval for the population test score mean. The random sample of 32 tests scores is shown next.

220 300 370 410 500 540 640 660 220 300 370 410 500 540 640 660 220 300 370 410 500 540 640 660 220 300 370 410 500 540 640 660

Level of Confidence = 95% = 1 - **α**

Level of Significance = **α** = 0.05

Sample Size = **n** = COUNT(Yellow Highlighted block of cells) = 32

Sample Mean = $\mathbf{x}_{\mathbf{avg}}$ = AVERAGE(Yellow Highlighted block of cells) = 455

Sample Standard Deviation = **s** = STDEV(Yellow Highlighted block of cells)= 149.8

Sample Standard Error = $\mathbf{s}_{\mathbf{xavg}}$ = **σ** / SQRT(**n**) ≈ **s** / SQRT(**n**)

$\mathbf{s}_{\mathbf{xavg}}$ = **σ** / SQRT(**n**) ≈ 149.8 / SQRT(32) = 26.5

Z Score$_{(1 - \alpha)}$ = Z Score$_{95\%}$ = NORMSINV (1 - **α**/2)
= NORMSINV (1 - 0.025) = NORMSINV (0.975) = 1.96

Width of Half the Confidence Interval = CONFIDENCE (**α, s, n**)

= CONFIDENCE (0.05, 149.6, 32) = 51.9

Also, equivalently:

Width of Half the Confidence Interval = Z $\text{Score}_{(1-\alpha)}$ * $\mathbf{s_{xavg}}$

= 1.96 * 26.5 = 51.9

Confidence Interval Boundaries = $\mathbf{x_{avg}}$ +/- Z $\text{Score}_{(1-\alpha)}$ * $\mathbf{s_{xavg}}$

= 455 +/- (1.96)*(26.5) = 455 +/- 51.9 = 403.1 to 506.9

Also, equivalently:

Confidence Interval Boundaries = $\mathbf{x_{avg}}$ +/- CONFIDENCE (**α, s, n**)

= 455 +/- 51.9 = 403.1 to 506.9

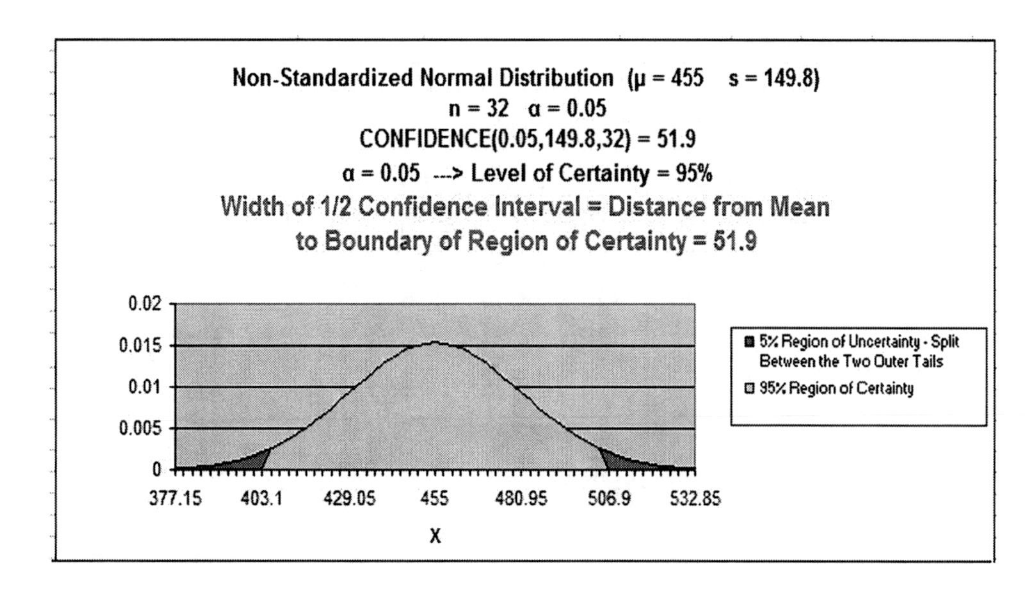

Problem 2: Calculate a Confidence Interval of Daily Sales Based Upon Sample Mean and Standard Deviation

Problem: Average daily demand for books sold in a small Barnes and Noble store is 455 books with a standard deviation of 200. This average and standard deviation are taken from sale data collected every day for a period of 60 days. What is the range that the true average daily book sales lies in with 95% certainty?

Level of Confidence = 95% = 1 - α

Level of Significance = α = 0.05

Sample Size = **n** = 60

Sample Mean = x_{avg} = 455

Sample Standard Deviation = **s** = 200

Sample Standard Error = s_{xavg} = σ / SQRT(**n**) ≈ **s** / SQRT(**n**)

s_{xavg} = σ / SQRT(**n**) ≈ 200 / SQRT(60) = 25.8

Z Score$_{(1-\alpha)}$ = Z Score$_{95\%}$ = NORMSINV (1 - α/2)

= NORMSINV (1 - 0.025) = NORMSINV (0.975) = 1.96

Width of Half the Confidence Interval = CONFIDENCE (**α**, **s**, **n**)

= CONFIDENCE (0.05, 200, 60) = 50.6
Also, equivalently:

Width of Half the Confidence Interval = Z $\text{Score}_{(1-\alpha)}$ * s_{xavg}

= 1.96 * 25.8 = 50.6

Confidence Interval Boundaries = x_{avg} +/- Z $\text{Score}_{(1-\alpha)}$ * s_{xavg}

= 455 +/- (1.96)*(25.8) = 455 +/- 50.6 = 404.4 to 505.6

Also, equivalently:

Confidence Interval Boundaries = x_{avg} +/- CONFIDENCE (**α**, **s**, **n**)

= 455 +/- 50.6 = 404.4 to 505.6

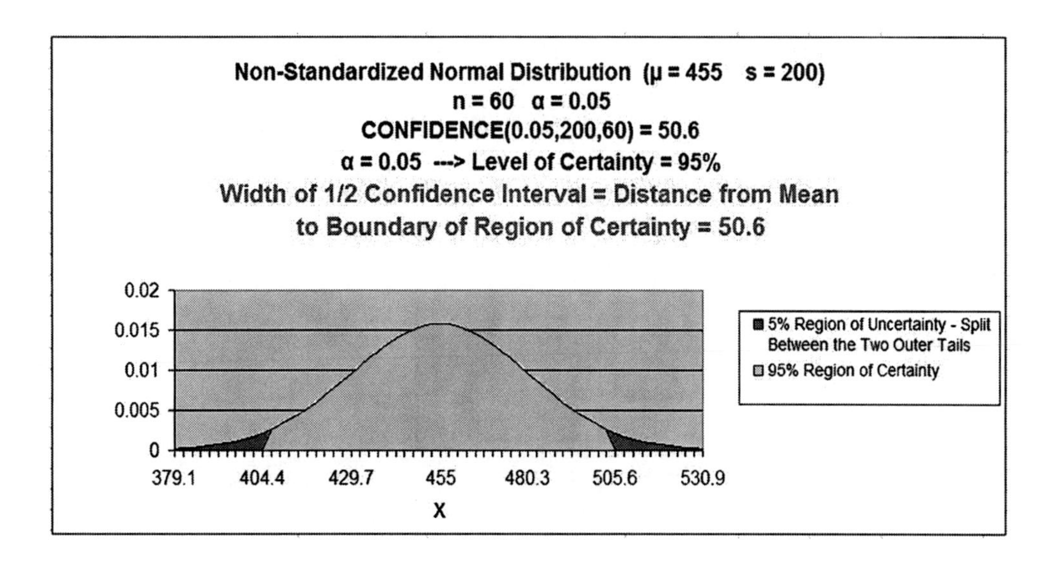

Problem 3: Calculate an Exact Range of 95% of Sales Based Upon the Upon the Population Mean and Standard Deviation

Problem: Average daily demand for books sold in a large Barnes and Noble store is 5,000 books with a standard deviation of 200. This average and standard deviation are taken from sale data collected every day for a period of 5 years. What is the range that 95% of the daily unit book sales falls in? The daily sales data is Normally distributed.

This problem is not a Confidence Interval problem. We do not need to create an estimate of the population mean (a Confidence Interval) because we know exactly what it is. We are given the population mean and population standard deviation.

We do need to know how the population is distributed in order to calculate the interval that contains 95% of all population data. Given that the population is Normally distributed, we simply need to map the region of this Normal curve that contains 95% of the total area and is centered about the mean as follows:

Population mean = μ = 5,000

Population Standard Deviation = σ = 200

Range Containing 95% of Sales Data = μ +/- [Z $\text{Score}_{95\%}$ * σ]

= 5,000 +/- [NORMSINV(0.975) * 200]
= 5,000 +/- 392
= 4,608 to 5,392

This is shown in the following diagram on the next page.

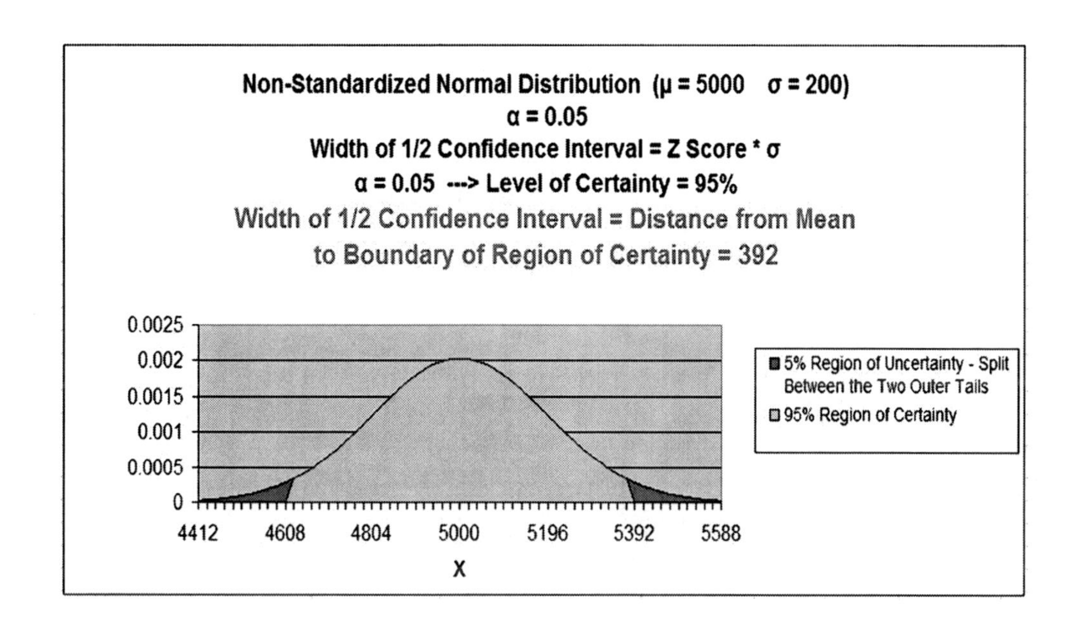
Non-Standardized Normal Distribution (μ = 5000 σ = 200)
α = 0.05
Width of 1/2 Confidence Interval = Z Score * σ
α = 0.05 ---> Level of Certainty = 95%
Width of 1/2 Confidence Interval = Distance from Mean
to Boundary of Region of Certainty = 392
0.0025
0.002
0.0015
0.001
0.0005
0
4412
4608
4804
5000
5196
5392
5588
X
5% Region of Uncertainty - Split Between the Two Outer Tails
95% Region of Certainty

Determining Minimum Sample Size (n) to Keep Confidence Interval of the Mean within a Certain Tolerance

The larger the sample size, the more accurate and tighter will be the prediction of a population's mean. Stated another way, the larger the sample size, the smaller will be the Confidence Interval of the population's mean. Width of the Confidence Interval is reduced when sample size is increased.

Quite often a population's mean needs to be estimated with some level of certainty to within plus or minus a specified tolerance. This specified tolerance is half the width of the Confidence Interval. Sample size directly affects the width of the Confidence Interval. The relationship between sample size and width of the Confidence Interval is shown as follows:

CONFIDENCE (α, s, n) = Width of half of the Confidence Interval

CONFIDENCE (α, s, n) = Z Score$_{(1-\alpha)}$ * s_{xavg}

Width of Half of the Confidence Interval = Z Score$_{(1-\alpha)}$ * s_{xavg}

Width of Half of the Confidence Interval = Z Score$_{(1-\alpha)}$ * σ / SQRT(**n**)

≈ Z Score$_{(1-\alpha)}$ * **s** / SQRT(**n**)

With algebraic manipulation of the above we have:

SQRT(**n**) = Z Score$_{(1-\alpha)}$ * σ / Width of Half of the Confidence Interval

n = [Z Score$_{(1-\alpha)}$]^2 * [σ]^2 / [Width of Half of the Confidence Interval]^2

Also, if we only have sample standard deviation, s, and not population standard deviation, σ :

n ≈ [Z Score$_{(1-\alpha)}$]^2 * [**s**]^2 / [Width of Half of the Confidence Interval]

Problem 4: Determine the Minimum Number of Sales Territories to Sample In Order To Limit the 95% Confidence Interval to a Certain Width

Problem: A national sales manager in charge of 5,000 similar territories ran a nationwide promotion. He then collected sales data from a random sample of the territories to evaluate sales increase. From the sample, the average sales increase per territory was $10,000 with a standard deviation of $500. How many territories would he have had to have sampled to be 95% sure that the actual nationwide average territory sales increase was no more than $50 different than average territory sales increase from the sample he took?

Level of Confidence = 95% = 1 - α
Level of Significance = α = 0.05

Sample Size = **n** = ?

Sample Mean = x_{avg} ------> Note this does not need to be known to solve this problem

Sample Standard Deviation = **s** = 500

Z Score(1 - α) = $Z_{Score95\%}$ = NORMSINV (1 - α/2)
= NORMSINV (1 - 0.025) = NORMSINV (0.975) = 1.96

Width of Half the Confidence Interval = 50

n ≈ [Z $\text{Score}_{(1-\alpha)}$]^2 * [**s**]^2 / [Width of Half of the Confidence Interval]^2
n ≈ [1.96]^2 * [500]^2 / [50]^2 = 384

The sales manager would have to sample at least 384 territories to be 95% certain that nationwide territory average was within +/- $50 of the sample territory average. Note that the 95% confidence interval is $10,000 +/- $50 and this interval has a width = $100 if sample size is 384.

Confidence Interval of a Population Proportion

Creating a Confidence Interval for a population's proportion is very similar to creating a Confidence Interval for a population's mean. The only real difference is how the standard error is calculated. Everything else is the same. The method of calculating a Confidence Interval for a population mean was covered in detail earlier in this module. First, the difference between using sampling to estimate a population mean and using sampling to estimate a population proportion will be explained below:

Mean Sampling vs. Proportion Sampling

What determines whether a mean is being estimated or a proportion is being estimated is the number of possible outcomes of each sample taken.

Proportion samples have only two possible outcomes. For example, if you are comparing the proportion of Republicans in two different cities, each sample has only two possible values; the person sampled either is a Republican or is not.

Mean samples have multiple possible outcomes. For example, if you are comparing the mean age of people in two different cities, each sample can have numerous values; the person sampled could be anywhere from 1 to 110 years old.

Below is a description of how to calculate a Confidence Interval for a population's proportion. Note that everything is almost the same as the calculation of the Confidence Interval for a mean, except sample standard error.

Levels of Confidence and Significance

Level of Significance, **α** ("alpha"), equals the maximum allowed percent of error. If the maximum allowed error is 5%, then $\alpha = 0.05$.

Level of Confidence is selected by the user. A 95% Level is the most common. A 95% Confidence Level would correspond to a 95% Confidence Interval of the Proportion. This would state that the actual population Proportion has a 95% probability of lying within the calculated interval. A 95% Confidence Level corresponds to a 5% Level of Significance, or $\alpha = 0.05$. The Confidence Level therefore equals $1 - \alpha$.

Population Proportion vs. Sample Proportion

Population Proportion = μ_p = **p** (This is what we are trying to estimate)

Sample Proportion = p_{avg}

Standard Deviation and Standard Error

Standard Deviation is not calculated during the creation of Confidence Interval for a population proportion.

Standard Error is an estimate of population Standard Deviation from data taken from a sample. Sample Standard Error will be an estimate taken from the sample proportion, p_{avg}, and sample size, n. This is the major difference between calculating a Confidence Interval for a proportion and for a mean. Binomial distribution rules apply to proportions because a proportion sample has only two possible outcomes, just like a binomial variable.

Sample Standard Error of a Proportion = σ_{pavg} = SQRT(**p** * **q** / **n**) ≈ s_{pavg}

Estimated Sample Standard Error of a Proportion = s_{pavg} = SQRT (p_{avg} * q_{avg} / **n**)

p = Population proportion - This is the unknown that will be estimated with a Confidence Interval

q = 1 - **p**

n = sample size

p_{avg} = Sample proportion

q_{avg} = 1 - p_{avg}

Region of Certainty vs. Region of Uncertainty

Region of Certainty is the area under the Normal curve that corresponds to the required Level of Confidence. If a 95% percent Level of Confidence is required, then the Region of Certainty will contain 95% of the area under the Normal curve. **The outer boundaries of the Region of Certainty will be the outer boundaries of the Confidence Interval**.

The Region of Certainty, and therefore the Confidence Interval, will be centered about the mean. Half of the Confidence Interval is on one side of the mean and half on the other side.

Region of Uncertainty is the area under the Normal curve that is outside of the Region of Certainty. Half of the Region of Uncertainty will exist in the right outer tail of the Normal curve and the other half in the left outer tail. This is similar to the concept of the "two-tailed test" that is used in Hypothesis testing in further sections of this course. The concepts of one and two-tailed testing are not used when calculating Confidence Intervals. Just remember that the Region of Certainty, and therefore the Confidence Interval, are always centered about the mean on the Normal curve.

Relationship Between Region of Certainty, Uncertainty, and Alpha

The Region of Uncertainty corresponds to α ("alpha").If α = 0.05, then that Region of Uncertainty contains 5% of the area under the Normal curve. Half of that area (2.5%) is in each outer tail. The 95% area centered about the mean will be the Region of Certainty. The outer boundaries of this Region of Certainty will be the outer boundaries of the 95% Confidence Interval. The Level of Confidence is 95% and the Level of Significance, or maximum error allowed, is 5%.

Z Score

Z Score is the number of Standard Errors from the mean to outer right boundary of the Region of Certainty (and therefore to the outer right boundary of the Confidence Interval). Standard Errors are used and not Standard Deviations because sample data is being used to calculate the Confidence Interval.

Z Score is calculated by the following Excel function:

Z Score(1-α) = NORMSINV (1 - α/2) - This will be discussed shortly.

Excel Functions Used When Calculating Confidence Interval for a Population Proportion

Note that Excel functions STDEV and AVERAGE are not used when working with proportions. The CONFIDENCE function is not used either.

COUNT

COUNT (Highlighted block of cells) = Sample size = **n**
----> Counts number of cells in highlighted block

NORMSINV

NORMSINV (1 - **α**/2) = Z $\text{Score}_{(1 - \alpha)}$

= Number of Standard Errors from mean to boundary of Confidence Interval. Note that (1 - **α**/2) = the entire area in the Normal curve to the left of outer right boundary of the Region of Certainty, or Confidence Interval. This includes the entire Region of Certainty and the half of the Region of Uncertainty that exists in the left tail.

For example:

Level of Confidence = 95% for a 95% Confidence Interval

Level of Significance = 5% (α = 0.05)

1 - α = 0.95 = 95%

Z $\text{Score}_{95\%}$ = NORMSINV (1 - α/2) = NORMSINV (1 - .05/2) = NORMSINV(1 - 0.025)

Z $\text{Score}_{95\%}$ = NORMSINV (0.975) = 1.96

The outer right boundary of the 95% Confidence Interval, and the Region of Certainty, is 1.96 Standard Errors from the mean. The left boundary is the same distance from the mean because the Confidence Interval is centered about the mean.

Formulas for Calculating Confidence Interval Boundaries from Sample Data for a Population Proportion

Confidence Interval Boundaries = Sample proportion +/- Z $\text{Score}_{(1-\alpha)}$ * Sample Standard Error

Confidence Interval Boundaries = $\mathbf{p_{avg}}$ +/- Z $\text{Score}_{(1-\alpha)}$ * $\mathbf{s_{pavg}}$

Sample Proportion = $\mathbf{p_{avg}}$

Z $\text{Score}_{(1-\alpha)}$ = NORMSINV (1 - α/2)

Sample Standard Error of a Proportion = $\boldsymbol{\sigma_{pavg}} \approx \mathbf{s_{pavg}}$ = SQRT ($\mathbf{p_{avg}}$ * $\mathbf{q_{avg}}$ / $\mathbf{n}$)

Sample size = $\mathbf{n}$ = COUNT (Highlighted block of cells containing samples)

Confidence Interval Boundaries = $\mathbf{p_{avg}}$ +/- Z $\text{Score}_{(1-\alpha)}$ * $\mathbf{s_{pavg}}$

Problem 5: Determine Confidence Interval of Shoppers Who Prefer to Pay By Credit Card Based Upon Sample Data

Problem : A random sample of 1,000 shoppers was taken. 70% preferred to pay with a credit card. 30% preferred to pay with cash. Determine the 95% Confidence Interval for the proportion of the general population that prefers to pay with a credit card.

Level of Confidence = 95% = 1 - α
Level of Significance = α = 0.05

Sample Size = n = 1,000

Sample Proportion = p_{avg} = 0.70

q_{avg} = 1 - p_{avg} = 0.30

Sample Standard Error of a Proportion = $\sigma_{pavg} \approx s_{pavg}$ = SQRT (p_{avg} * q_{avg} / **n**)

s_{pavg} = SQRT (0.70 * 0.30 / 1,000) = 0.014

Z Score$_{(1-\alpha)}$ = Z Score95% = NORMSINV (1 - **α**/2)
= NORMSINV (1 - 0.025) = NORMSINV (0.975) = 1.96

Width of Half the Confidence Interval = Z Score$_{(1-\alpha)}$ * s_{pavg}
= 1.96 * 0.014 = 0.0274

Confidence Interval Boundaries = p_{avg} +/- Z Score$_{(1-\alpha)}$ * s_{pavg}
= 0.70 +/- (1.96) * (0.014)
= 0.70 +/- 0.0274 = 0.6726 to 0.7274 = 67.26% to 72.74%

This is shown in the following diagram on the next page:

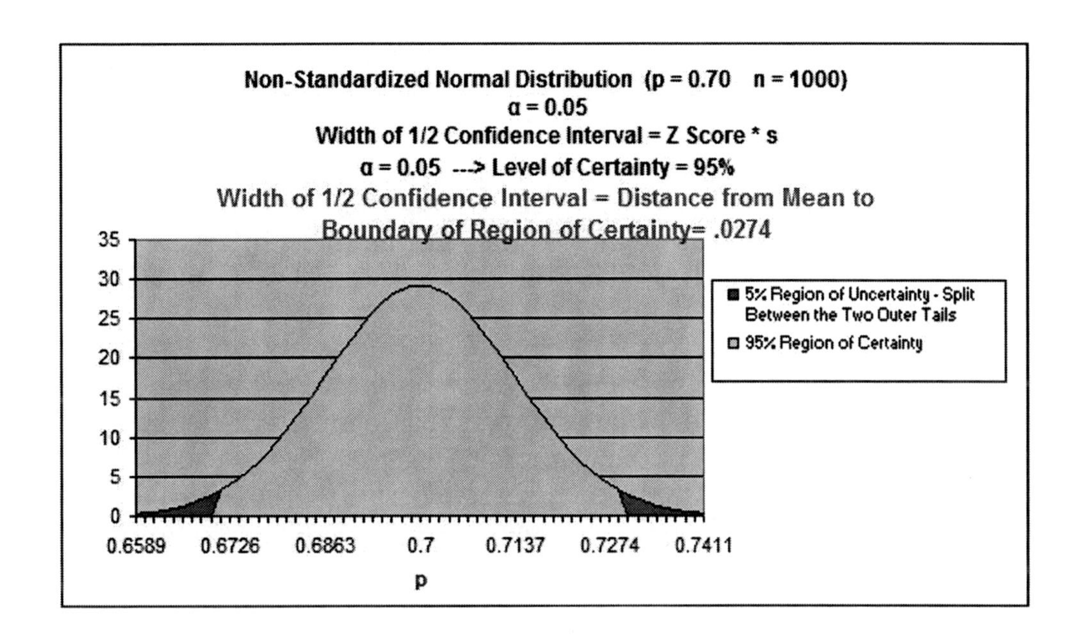
Non-Standardized Normal Distribution (p = 0.70 n = 1000)
α = 0.05
Width of 1/2 Confidence Interval = Z Score * s
α = 0.05 ---> Level of Certainty = 95%
Width of 1/2 Confidence Interval = Distance from Mean to Boundary of Region of Certainty= .0274
35
30
25
20
15
10
5
0
0.6589
0.6726
0.6863
0.7
0.7137
0.7274
0.7411
p
5% Region of Uncertainty - Split Between the Two Outer Tails
95% Region of Certainty

Determining Minimum Sample Size (n) to Keep Confidence Interval of the Proportion within a Certain Tolerance

The larger the sample size, the more accurate and tighter will be the prediction of a population's mean. Stated another way, the larger the sample size, the smaller will be the Confidence Interval of the population's mean. Width of the Confidence Interval is reduced when sample size is increased.

Quite often a population's mean needs to be estimated with some level of certainty to within plus or minus a specified tolerance. This specified tolerance is half the width of the confidence interval. Sample size directly affects the width of the Confidence Interval. The relationship between sample size and width of the Confidence Interval is shown as follows:

Width of Half the Confidence Interval = Z $\text{Score}_{(1-\alpha)}$ * s_{pavg}

s_{pavg} = SQRT (p_{avg} * q_{avg} / **n**)

Width of Half the Confidence Interval = Z $\text{Score}_{(1-\alpha)}$ * SQRT (p_{avg} * q_{avg} / **n**)

[Width of Half the Confidence Interval]^2 = [Z $\text{Score}_{(1-\alpha)}$]^2 * (p_{avg} * q_{avg} / **n**)

n = [Z Score**(1-α)**]^2 * (p_{avg} * q_{avg}) / **[Width of Half the Confidence Interval]**^{2}

Problem 6: Determine the Minimum Sample Size of Voters to be 95% Certain that the Population Proportion is no more than 1% Different from Sample Proportion

Problem: A random survey was conducted in one city to learn voting preferences. 40% of voters surveyed said they would vote Republican. 60% of the voters surveyed said they would vote Democrat. Determine the minimum number of voters that had to be surveyed to be 95% certain that the results were accurate within +/- 1%.

Level of Confidence = 95% = 1 - α

Level of Significance = α = 0.05

$\mathbf{p_{avg}}$ = 0.40

$\mathbf{q_{avg}}$ = 1 - $\mathbf{p_{avg}}$ = 0.60

Width of Half the Confidence Interval = 0.01 ---> (1%)

Z Score$_{(1-\alpha)}$ = Z Score$_{95\%}$ = NORMSINV (1 - $\alpha/2$)
= NORMSINV (1 - 0.025) = NORMSINV (0.975) = 1.96

n = [Z Score$_{(1-\alpha)}$]**2** * ($\mathbf{p_{avg}}$ * $\mathbf{q_{avg}}$) / [Width of Half the Confidence Interval]^2

n = [1.96]^2 * (0.40 * 0.60) / [0.01]^2 = 9,220

At least 9,220 random voters had to be surveyed to be 95% certain that the population proportion is no more than 1% different from the sample.

Hand Calculation of Confidence Interval Problems

Go To
http://excelmasterseries.com/Excel_Statistical_Master/Confidence-Interval.php

To View How To Solve Confidence Interval Problems By Hand (No Excel)

(Is Your Internet Connection Turned On ?)

You'll Quickly See Why You Always Want To Use Excel To Solve Statistical Problems !

Chapter 8 - Hypothesis Testing of Means

Basic Explanation of Hypothesis Testing

Hypothesis testing for change is definitely one of the most useful statistical tools for the business manager. Quite often we need to determine within a small possibility of error whether something measurable has changed. For example, you have implemented a new advertising campaign and you want to determine if sales really have improved. Hypothesis testing can be used in a wide variety of situations where you have Before and After data and you want to determine if real change has occurred. Hypothesis testing is also a great tool to statistically verify whether two groups have the same mean or proportion of something.

Hypothesis testing involves creating two separate Hypotheses and then testing to see which one applies. The two hypotheses are the Null Hypothesis and the Alternate Hypothesis. The Null Hypothesis, often referred to as Ho, usually states that both means are the same or that no change has occurred. The Alternate Hypothesis, H1, states the mean has changed. In other words, the new mean is statistically different from the old mean.

Hypothesis testing uses sample data to verify either the Null or Alternate Hypothesis about a population. The most important goal of Hypothesis testing is to verify the correct hypothesis about a population within a specified degree of certainty. For example, you have run a new ad campaign for all of your dealers and you want to be at least 95% sure whether sales have increased throughout your entire dealer network based upon a random sample of dealer sales results taken before and after the ad campaign.

The Four-Step Method to Solving ALL Hypothesis Test Problems

There are many types of Hypothesis tests but fortunately they can all be solved in the same general way using this four-step method:

Step 1 - Create the Null Hypothesis and the Alternate Hypothesis

Step 2 - Map the Normal Curve

Step 3 - Map the Region of Certainty

Step 4 - Perform the Critical Value Test or the p-Value Test

We will cover these steps in more detail later but ALL Hypothesis testing can be done with this four-step method. All of the different Hypothesis test problems presented in this course are solved with the four-step method.

The Four Ways of Classifying ALL Hypothesis Test Problems

1) Mean Testing vs. Proportion Testing

The basic objective of Hypothesis testing is to determine whether the mean or proportion within one group is statistically the same as the mean or proportion within another group. What determines whether a mean is being tested or a proportion is being tested is the number of possible outcomes of each sample taken.

Proportion test samples have only two possible outcomes. For example, if you are comparing the proportion of Republicans in two different cities, each sample has only two possible values; the person sampled either is a Republican or is not.

Mean test samples have multiple possible outcomes. For example, if you are comparing the mean age of people in two different cities, each sample can have numerous values; the person sampled could be anywhere from 1 to 110 years old.

Hypothesis tests of mean are computed the same as Hypothesis tests of proportion in every way except one: the calculation of the Sample Standard Error. This difference is significant enough that these two types of Hypothesis tests are analyzed in separate modules of this course.

2) One-Tailed vs. Two-Tailed Testing

All Hypothesis tests are either one-tailed or two-tailed. The number of tails depends on whether the Hypothesis test can determine if the mean or proportion in one sampled group is merely different than in another group or whether it is different in one direction (is either larger or smaller) than in another group

A one-tailed test is used to determine whether a mean or proportion is different in one direction than another mean or proportion, not that it is merely different.

A one-tailed test would be used to determine if the proportion of Republicans in one city is larger than the proportion of Republicans in another city. Another one-tailed test would be used to determine if the mean age of people in one city is less than the mean age of people in another city.

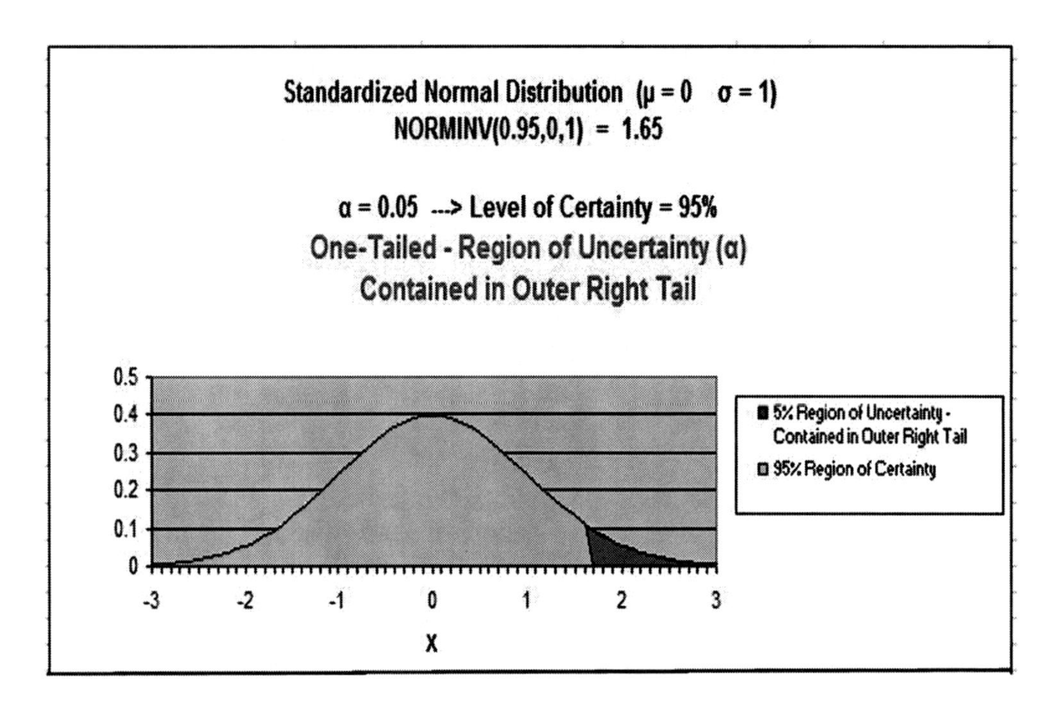
Standardized Normal Distribution (μ = 0 σ = 1)
NORMINV(0.95,0,1) = 1.65
α = 0.05 ---> Level of Certainty = 95%
One-Tailed - Region of Uncertainty (α)
Contained in Outer Right Tail
0.5
0.4
0.3
0.2
0.1
0
-3
-2
-1
0
1
2
3
X
5% Region of Uncertainty - Contained in Outer Right Tail
95% Region of Certainty

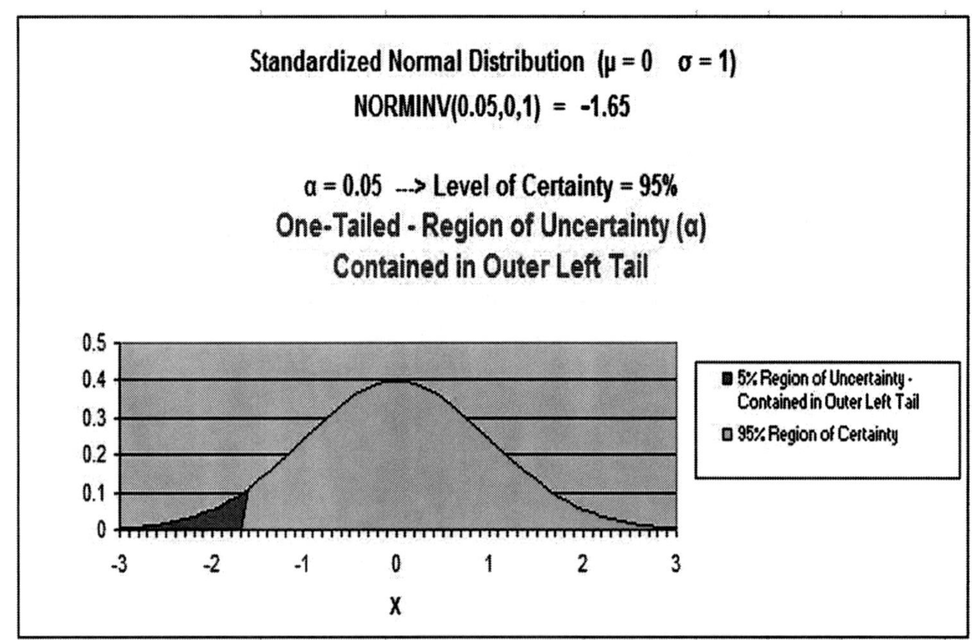
Standardized Normal Distribution (μ = 0 σ = 1)
NORMINV(0.05,0,1) = -1.65
α = 0.05 ---> Level of Certainty = 95%
One-Tailed - Region of Uncertainty (α)
Contained in Outer Left Tail
0.5
0.4
0.3
0.2
0.1
0
-3
-2
-1
0
1
2
3
X
5% Region of Uncertainty - Contained in Outer Left Tail
95% Region of Certainty

Two-tailed tests are used to determine whether a mean or proportion is merely different than another mean or proportion. The direction of the difference is not a factor in a two-tailed test, as it is in a one-tailed test.

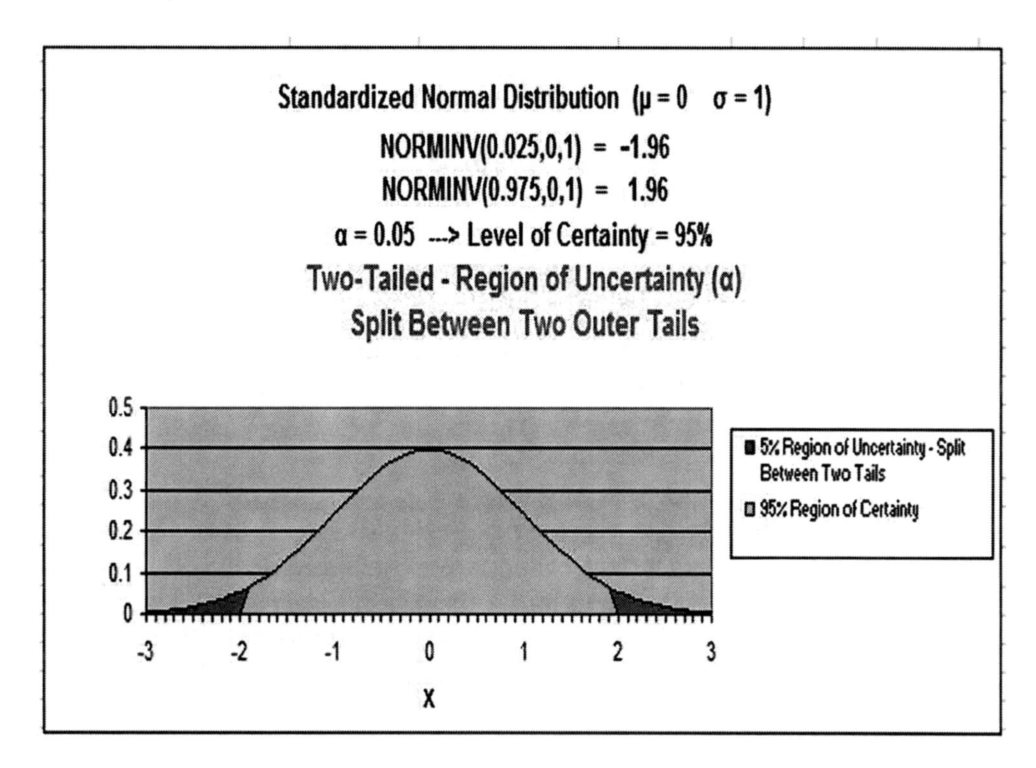

3) One-Sample vs. Two-Sample Testing

Hypothesis testing fundamentally involves comparing one mean or proportion with another mean or proportion. Whether you need to take one sample or two samples depends upon whether you already have original or "Before" data available.

One-sample testing is performed if original or "Before" comparison data is already in place at the start of the test. A one-sample hypothesis test is normally performed to determine whether the original or "Before" data is still valid or whether something has changed.

Two-sample testing is performed if no "Before" data is available or if a comparison is being made but no data is available on either side.

To summarize - if data is available for one of the means or proportions being compared, then only one sample is needed for data collection of the other mean or proportion being compared. If no data is available, then two samples must be taken - one for each mean or proportion being compared.

4) Unpaired Data Testing vs. Paired Data Testing

Most Hypothesis testing uses unpaired data testing. Whether data is paired or unpaired depends on whether both samples were collected from the same objects or not.

For data to be Paired data, both samples must have been collected from exactly the same objects. An example of this would be "Before" and "After" sales data taken from the same individual dealers to test whether an ad campaign had increased sales. The dealers are randomly chosen and each dealer provides a single set of "Before" and "After" data. Hypothesis testing is performed on the set of differences between "Before" and "After" numbers for each data pair. Mean testing, but not proportion testing, can be performed using paired data.

Unpaired data samples are group samples collected independently of each other. Groups of unpaired data are treated independently of each other. Separate means and standard errors are calculated from each group. Hypothesis testing is then performed to compare the means or proportions of the two separate groups. The majority of Hypothesis testing is performed using Unpaired data.

Detailed Description of the Four-Step Method for Solving Mean Testing Problems

The four-step method is used to solve all Hypothesis testing problems. This course module will discuss solving tests of mean. The next module of this course will cover solving tests of proportion. Both types of tests are solved in the same general way. The main difference between solving for mean and proportion is the calculation of sample standard error. Below is a detailed description of each of the four steps required to solve a Hypothesis test of mean:

Initial Steps

Before solving a hypothesis test problem, you must classify the problem type and lay out the information given in the problem properly.

Problem Classification

Problem Classification: Select the proper choice of each of the four ways that a Hypothesis problem is classified as follows:

1) Mean Testing vs. Proportion Testing

• Proportion test samples have only two possible outcomes.

• Mean test samples have multiple possible outcomes.

2) One-Tailed vs. Two-Tailed Testing

• Two-tailed tests determine whether two means are merely different.

• One-tailed tests determine whether one mean is different in one direction.

3) One-Sample vs. Two Sample Testing

• One sample is taken if original or "Before" comparison data is available.

• Two samples are taken if no comparison data is available.

4) Unpaired Data Testing vs. Paired Data Testing

• Paired data testing can be performed if "Before" and "After" data are collected from the same objects. Mean testing can be performed on paired data - Proportion testing cannot.

• Unpaired data testing is performed on data collected in groups.

Information Layout

Information Layout: Listing the given information in a problem properly greatly expedites problem solving. Here is a list of given information that needs to be laid out before solving:

1) Level of Significance

The Level of Significance, α, is equivalent to the maximum possibility of error. The Level of Significance can also be derived from the Required Level of Certainty.

2) Existing Comparison Data

The includes an existing population that is being verified or the "Before" mean that will be compared to the "After" mean. The existing mean data will normally also contain standard deviation or standard error information.

3) Comparison Sample Data

Any sample data will include the sample average mean and sample size. Sample data may also include sample standard deviation or sample standard error.

Sample Standard Error = **(**Standard Deviation / SQRT[Sample size] **)**

The Four Steps to Hypothesis Testing

Descriptions of each of the four steps are presented here but if you would like to see these steps being applied to all of the different types of mean Hypothesis tests, all problems completed in this module use these four steps. Proportion Hypothesis testing using these four steps is shown and explained in the next module of this course. Directly below is a description of Mean Hypothesis testing.

Step 1 - Create the Null and Alternate Hypotheses

The **Null Hypothesis states that both means are the same.**

For a two-sample test, the means of both samples, x1avg and x2avg, are being compared. The Null Hypothesis states that they are both equal as follows:

Null Hypothesis, H_0 -----> $\mathbf{x_{1avg}} - \mathbf{x_{2avg}} = 0$

For a one-sample test, the mean of the sample taken, $\mathbf{x_{avg}}$, is **compared to the Constant** that the original or "Before" mean, $\boldsymbol{\mu}$, was measured to be. The Null Hypothesis states that both are equivalent as follows:

Null Hypothesis, H_0 ----> $\mathbf{x_{avg}}$ = **Constant**

The **Alternate Hypothesis, H1, states that both means are different**.

A **two-tailed test** states that the two means are merely different as follows:

- One-sample, two-tailed test Alternate hypothesis

H_1 ----> $\mathbf{x_{avg}} \neq$ **Constant**

- Two-sample, two-tailed test Alternate hypothesis

H_1 ----> $\mathbf{x_{1avg} - x_{2avg} \neq Constant}$

A **one-tailed test** states that the two means are different in one direction as follows:

- One-sample, one-tailed test Alternate hypothesis

H_1 ----> $\mathbf{x_{avg} > Constant}$ **OR** $\mathbf{x_{avg} < Constant}$

- Two-sample, one-tailed test Alternate hypothesis

H_1 ----> $\mathbf{x1_{avg} - x2_{avg} > Constant}$ **OR** $\mathbf{x_{1avg} - x_{2avg} < Constant}$

To summarize:

One-tailed test ----> (Value of variable) **is greater** than OR **is less than** (Constant)

Two-tailed test ----> (Value of variable) **does not equal** (Constant)

Step 2 - Map the Normal Curve

We now create a Normal curve showing a distribution of the same variable that is used by the Null Hypothesis, which is $\mathbf{x_{avg}}$ or ($\mathbf{x_{1avg}}$ - $\mathbf{x_{2avg}}$)

The mean of this Normal curve will occur at the same value of the distributed variable as stated in the Null Hypothesis.

Since the Null Hypothesis states that either ($\mathbf{x_{avg}}$) **OR** ($\mathbf{x_{1avg}}$ - $\mathbf{x_{2avg}}$) = **Constant** the Normal curve will map the distribution of the variable ($\mathbf{x_{avg}}$) **OR** ($\mathbf{x_{1avg}}$ - $\mathbf{x_{2avg}}$) with a mean equal to the **Constant**.

This Normal curve will have a standard error whose calculation depends on whether the mean test uses one sample or two samples as follows:

One Sample Test

Sample Standard Error = $\mathbf{s_{xavg}}$ = $\boldsymbol{\sigma}$ / SQRT(**n**)

$\boldsymbol{\sigma}$ = Population original or "Before" standard deviation
n = sample size

Two Sample Test

Standard Error = $\mathbf{s_{(x1avg - x2avg)}}$ = SQRT [(s_1^2) / n1 + (s_2^2) / n2]

s1 = Standard deviation of Sample 1
s2 = Standard deviation of Sample 2
n1 & n2 are sample sizes

Excel Note - Standard deviation is calculated by the Excel function STDEV( )

Step 3 - Map the Region of Certainty

The **Region of Certainty** is the percentage of area under the Normal curve that corresponds with the degree of certainty required by the problem. For example, if the problem requires at least 95% certainty, the Region of Certainty will contain 95% of the area under the Normal curve.

The remainder of the area will be contained in the Region of Uncertainty. The **Region of Uncertainty** is the percentage of area under the Normal curve that corresponds with the maximum allowed possibility of error. For example, if the problem requires at least 95% certainty, then the max chance of error is 5%. The Region of Uncertainty would therefore contain 5% of the total area under the Normal curve.

The area in the Region of Uncertainty corresponds to **α (alpha)**. For example, if the problem allows a max chance of error of 5%, then 5% of the total area will be contained in the Region of Uncertainty and α = 0.05.

Mapping the Region of Uncertainty for a Two-Tailed Test:

The **Region of Uncertainty** will be split between **both outer tails**. Each outer tail will contain α/2 of the total area under the Normal curve.

Mapping the Region of Uncertainty for a One-Tailed Test:

The **Region of Uncertainty** will be contained in **one outer tail**. That outer tail will contain α of the total area under the Normal curve.

The Alternate Hypothesis determines whether the Region of Uncertainty is contained in the left or the right outer tail.

The **Region of Uncertainty** will be contained in the **right outer tail** If: Alternate Hypothesis states ---> (Value of variable) **is greater than** (Constant)

The **Region of Uncertainty** will be contained in the **left outer tail** If: Alternate Hypothesis states ---> (Value of variable) **is less than** (Constant)

Mapping the Region of Certainty for a Two-Tailed Test

Mapping the Region of Certainty means calculating how far one or both boundaries of the Region of Certainty are from the Normal curve's mean in its center. The Region of Certainty has only one outer boundary in a one-tailed test but has two outer boundaries in a two-tailed test.

The Region of Certainty for a two-tailed test is located in the middle of the Normal curve in between the equal Regions of Uncertainty in each outer tail. The left and right outer boundaries of the Region of Certainty are both the same distance from the Normal curve mean. The number of standard errors that either of these two outer boundaries is from the mean is calculated as follows in this example:

For a two-tailed test with a required level of certainty of **95%**, the number of standard errors from the Normal curve mean to either of the outer boundaries of the Region of Certainty is calculated in Excel as follows:

$Z_{95\%,2\text{-tailed}}$ = NORMSINV(1 - α/2) = NORMSINV(**0.975**) = 1.96

Excel Note - NORMSINV(x) = The number of standard errors from the Normal curve mean to a point right of the Normal curve mean at which x percent of the area under the Normal curve will be to the left of that point. For a two-tailed test, x equals the Level of Certainty plus 1/2 the Level of Uncertainty.

This occurs because the area under the curve to the left of x will contain the entire Region of Certainty and the 1/2 part of the Region of Uncertainty that exists in the outer left tail.

The Region of Certainty extends to the left and to the right of the Normal curve mean by 1.96 standard errors.

If, for example, one standard error = $\mathbf{s_{xavg}}$ = 0.51, then:

1.96 standard errors = (1.96) * (0.51) = 0.9996

The outer boundaries of the Region of Certainty have the values:
$= \mathbf{\mu} +/- \mathbf{Z_{95\%,2\text{-tailed}}} * \mathbf{s_{xavg}}$

If the Normal curve mean, which is established in the Null Hypothesis equals 0, then the outer boundaries of the Region of Certainty have the values 0 +/- (1.96) * (0.51) = +/- 0.9996

These points (-0.9996 and 0.9996) are 1.96 standard errors from the Normal curve mean of $\mathbf{x_{avg}}$ = 0

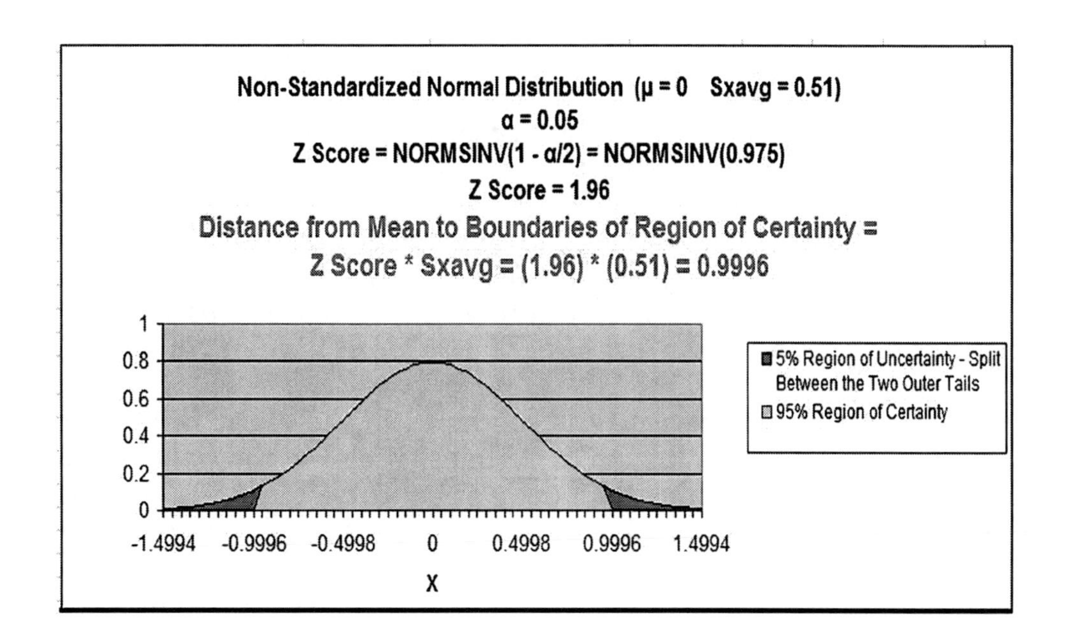

Mapping the Region of Certainty for a One-Tailed Test

The Region of Certainty for a one-tailed test has only one outer boundary. The entire Region of Uncertainty for the one-tailed test is contained in only one outer tail. The remainder of the area under the Normal curve makes up the Region of Certainty.

The Excel function for calculating the distance from the Normal curve mean to the one boundary of the Region of Certainty is
shown in this example:

For a one-tailed test with a required level of certainty of **95%**, the number of standard errors from the Normal curve mean to the outer boundary of the region of Certainty is calculated in Excel as follows:

$Z_{95\%,1\text{-tailed}}$ = NORMSINV(1 - α) / NORMSINV(**0.95**) = 1.65

Excel Note - NORMSINV(x) = The number of standard errors from the Normal curve mean to a point right of the Normal curve mean at which x percent of the area under the Normal curve will be to the left of that point. For a one-tailed test, x equals the Level of Certainty.

Important Excel Note ---> The above Excel function calculates the number of standard errors from the mean to the boundary of the Region of Certainty regardless of whether that boundary is on the left side or the right side of the mean.

If the **Region of Uncertainty** is contained in the **right outer tail:** then the boundary of the Region of Certainty is also in the right tail and therefore has a greater value than the Normal curve mean, **μ**. The value of boundary for the 95% Region of Certainty in the right tail for a one-tailed test is:

$\mu + Z_{95\%,1\text{-tailed}} * s_{xavg}$

The 95% Region of Certainty extends to the right of the Normal curve mean by 1.65 standard errors for this one-tailed test, shown as follows.

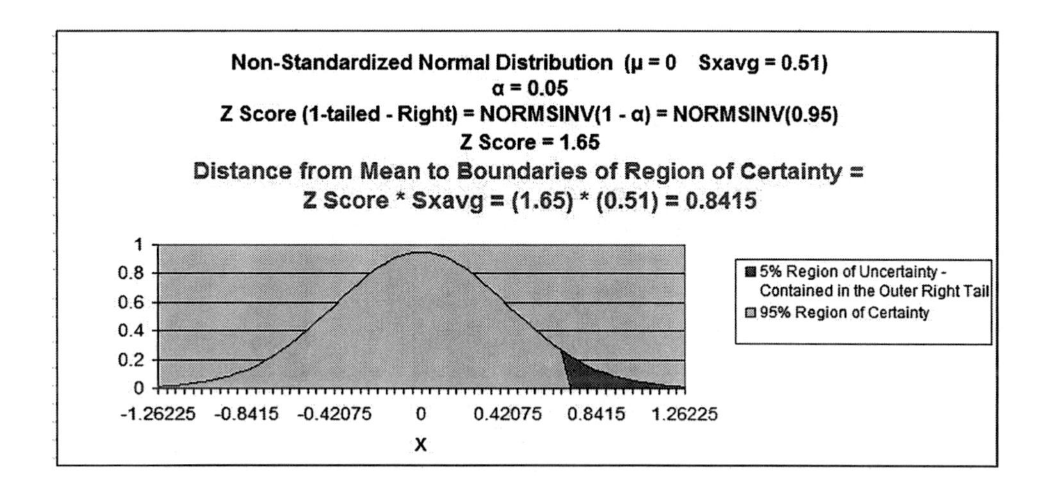

If, for example, one standard error = $\mathbf{s_{xavg}}$ = 0.51, then:

1.65 standard errors = (1.65) * (0.51) = 0.8415

The one and only outer boundary of this 95% Region of Certainty has the value: $\boldsymbol{\mu} + \mathbf{Z_{95\%,one\text{-}tailed}} * \mathbf{s_{xavg}}$

If the Normal curve mean, which is established in the Null Hypothesis equals 0, then the outer boundary of the Region of Certainty has the value: 0 + (1.65) * (0.51) = 0.8415

This point (0.8415) is 1.65 standard errors to the right of the Normal curve mean of $\mathbf{x_{avg}}$ = 0

If the Region of Uncertainty is contained in the left outer tail: then the boundary of the Region of Certainty is also in the left tail and therefore has a lower value than the Normal curve mean, **μ**. The value of boundary for the Region of Certainty in the left tail for a one-tailed test is:

$\boldsymbol{\mu} - \mathbf{Z_{95\%,1\text{-}tailed}} * \mathbf{s_{xavg}}$

The Region of Certainty extends to the left of the Normal curve mean by 1.65 standard errors.

If, for example, one standard error = $\mathbf{s_{xavg}}$ = 0.51, then:

1.65 standard errors = (1.65) * (0.51) = 0.8415

The one and only outer boundary of this Region of Certainty has the values = $\boldsymbol{\mu}$ - $\mathbf{Z_{95\%,1\text{-tailed}}}$ * $\mathbf{s_{xavg}}$

If the Normal curve mean, which is established in the Null Hypothesis equals 0, then the outer boundary of the Region of Certainty has the value: 0 - (1.65) * (0.51) = -0.8415

This point (-0.8415) is 1.65 standard errors to the left of the Normal curve mean of $\mathbf{x_{avg}}$ = 0

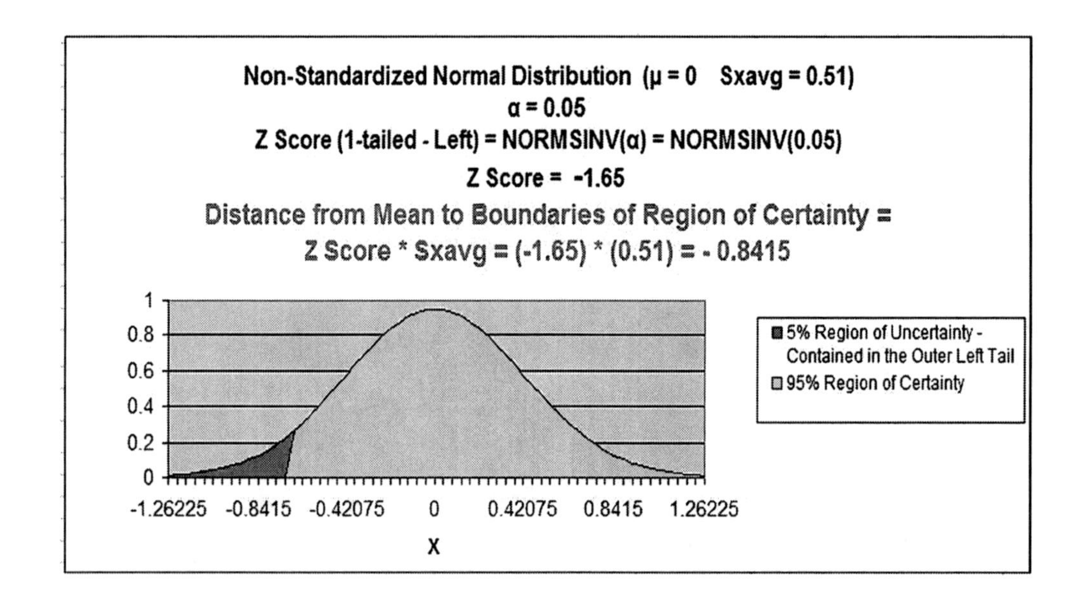

Step 4 - Perform Critical Value and p-Value Tests

a) Critical Value Test

The Critical Value Test is the final test to determine whether to reject or fail to reject the Null Hypothesis. The p Value Test, described on the next page, is an equivalent alternative to the Critical Value Test.

The Critical Value test tells whether the value of the actual variable, xavg, falls inside or outside of the **Critical Value**, which is the **boundary between the Region of Certainty and the Region of Uncertainty**.

If the actual value of the distributed variable, (x_{avg}) or (x_{1avg} - x_{2avg}), falls within the Region of Certainty, the Null Hypothesis is not rejected.

If the actual value of the distributed variable, (x_{avg}) or (x_{1avg} - x_{2avg}), falls outside of the Region of Certainty and, therefore, into the Region of Uncertainty, the Null Hypothesis is rejected and the Alternate Hypothesis is accepted.

The Critical Value test is much easier to implement with Excel than the p Value test. You will believe it after you review the p Value test next.

b) p Value Test

The p Value Test is an equivalent alternative to the Critical Value Test and also tells whether to reject or fail to reject the Null Hypothesis.

The p Value equals the percentage of area under the Normal curve that is in the tail outside of the actual value of the variable (x_{avg}) or (x_{1avg} - x_{2avg}).

For a one-tailed test, if the p Value is larger than α, the Null Hypothesis is not rejected. For a two-tailed test, if the p Value is larger than $\alpha/2$, the Null Hypothesis is not rejected.

For a one-tailed test, the Region of Uncertainty is contained entirely in one tail. Therefore the curve area contained by the Region of Uncertainty in that tail equals **α**,

For a two-tailed test, the Region of Uncertainty is split between both tails. Therefore the curve area contained by the Region of Uncertainty in that tail equals **α**/2,

There are two possibilities for calculating the p Value. Each possibility depends on whether the actual value of the distributed variable ($\mathbf{x_{avg}}$) or ($\mathbf{x_{1avg}}$ - $\mathbf{x_{2avg}}$) is greater or less than the Normal curve mean, which is established by the Null Hypothesis.

Actual value of the variable is greater than (to the right of) the mean.

p Value_{xavg} = **1** - NORMSDIST(**[** $\mathbf{x_{avg}}$ - $\boldsymbol{\mu}$ **]** / $\mathbf{s_{xavg}}$)

(**[**$\mathbf{x_{avg}}$ - $\boldsymbol{\mu}$**]** / $\mathbf{s_{xavg}}$) = number of standard errors that $\mathbf{x_{avg}}$ is from the mean

Excel note - NORMSDIST(x) calculates the total area under the Normal curve to the left of the point that is x standard errors to the right of the Normal curve mean. If x is to the right of the mean, x will be positive and NORMSDIST(x) will be greater than 0.50. If x is to the left of the mean, x will be negative and NORMSDIST(x) will be less than 0.50.

Actual value of the variable is less than (to the left of) the mean.

p Value_{xavg} = NORMSDIST(**[**$\mathbf{x_{avg}}$ - $\boldsymbol{\mu}$**]** / $\mathbf{s_{xavg}}$)

(**[**$\mathbf{x_{avg}}$ - $\boldsymbol{\mu}$**]** / $\mathbf{s_{xavg}}$) = number of standard errors the $\mathbf{x_{avg}}$ is from the mean

Excel note - NORMSDIST(x) calculates the total area under the Normal curve to the left of the point that is x standard errors to the right of the Normal curve mean.

For a two-tailed test---> When the p Value is greater than α/2, the actual value of the distributed variable falls inside the Region of Certainty and the Null Hypothesis is not rejected.

For a one-tailed test---> When the p Value is greater than α, the actual value of the distributed variable falls inside the Region of Certainty and the Null Hypothesis is not rejected.

The p Value test is better understood by examining the example problems following. As mentioned before, the Critical Value test is equivalent to the p Value test but is much easier to implement in Excel.

Type 1 and Type 2 Errors

Type 1 Error occurs if the null hypothesis is incorrectly rejected when it is actually true. In other words, it is incorrectly believed that a parameter changed when it did not.

Type 2 Error occurs if the null hypothesis is incorrectly not rejected when it is actually false. In other words, it is incorrectly believed that a parameter did not change but it did.

Problems

Problem 1 - Two-Tailed, One-Sample, Unpaired Hypothesis Test of Mean - Testing a Manufacturer's Claim of Average Product Thickness

Problem: A manufacturer claims that the average thickness of metal sheets is 15 mls. and that the population standard deviation, σ, is 0.1 mls. 50 sheets are sampled having a sample mean of 14.982 mls. At the 0.05 Significance Level (95% Level of Certainty) state whether the manufacturer's claim that the average thickness of 15 mls. is reliable.

We know that this is a **test of mean** and not proportion because **each individual sample taken can have a wide range of values**: Any sheet thickness measurement from 14.90 to 15.10 is probably reasonable.

We know that this is a **two-tailed** test because we are trying to **determine if the "Before Data" and "After Data" means are merely different**, not whether one mean is larger or smaller than the other.

We know that only **one sample** needs to be taken because the initial population data is already available.

This is **unpaired data** because groups are sampled independently.

"Before Data"

μ = "Before Data" mean = 15

σ = "Before Data" population standard deviation = 0.1

"After Data"

X_{avg} = "After Data" sample average = 14.982

n = "After Data" Sample size = 50

α = Level of Significance = 0.05 ---> 5% Max chance of error ---> 95% Level of Certainty Required

This problem can be solved using the standard four-step method for Hypothesis testing.

Step 1 - Create the Null and Alternate Hypotheses

The Null Hypothesis normally states that both means are the same. If the "Before Data" population mean, μ, equals the "After Data" sample mean, $\mathbf{x_{avg}}$, then $\mathbf{x_{avg}} = \boldsymbol{\mu} = 15$

The Null Hypothesis states that both means are the same, which is equivalent to:

The Null Hypothesis, which states that $\mathbf{x_{avg}}$ is the same as $\boldsymbol{\mu}$ (which is 15), is as follows:

Null Hypothesis, H_o ----> $\mathbf{x_{avg}} = 15$

The Alternate Hypothesis states that the means are different, which is equivalent to:

The Alternate Hypothesis, which states that $\mathbf{x_{avg}}$ is different than $\boldsymbol{\mu}$ (which is 15, is as follows:

Alternate Hypothesis, H**1** ----> $\mathbf{x_{avg}} \neq 15$

For this two-tailed test, the Alternative Hypothesis states that the value of the distributed variable xavg does not equal the value of 15 stated in the Null Hypothesis,

The **Region of Uncertainty** will be split between **both outer tails**.

Note - the Alternative Hypothesis determines whether the Hypothesis test is a one-tailed test or a two-tailed test as follows:

One-tailed test ----> (Value of variable) **is greater than** OR **is less than** (Constant)

Two-tailed test ----> (Value of variable) **does not equal** (Constant)

Step 2 - Map the Normal Curve

We now create a Normal curve showing a distribution of the same variable that is used by the Null Hypothesis, which is $\mathbf{x_{avg}}$.

The mean of this Normal curve will occur at the same value of the distributed variable as stated in the Null Hypothesis.

Since the Null Hypothesis states that $\mathbf{x_{avg}}$ = 15, the Normal curve will map the distribution of the variable $\mathbf{x_{avg}}$ with a mean of $\mathbf{x_{avg}}$ = 15

****

This Normal curve will have a standard error that is calculated as the standard error of a sample taken from a population is normally calculated, as follows:

Sample Standard Error = $\mathbf{s_{xavg}}$ = $\boldsymbol{\sigma}$ / SQRT(**n**) = 0.1 / SQRT(50) = 0.014

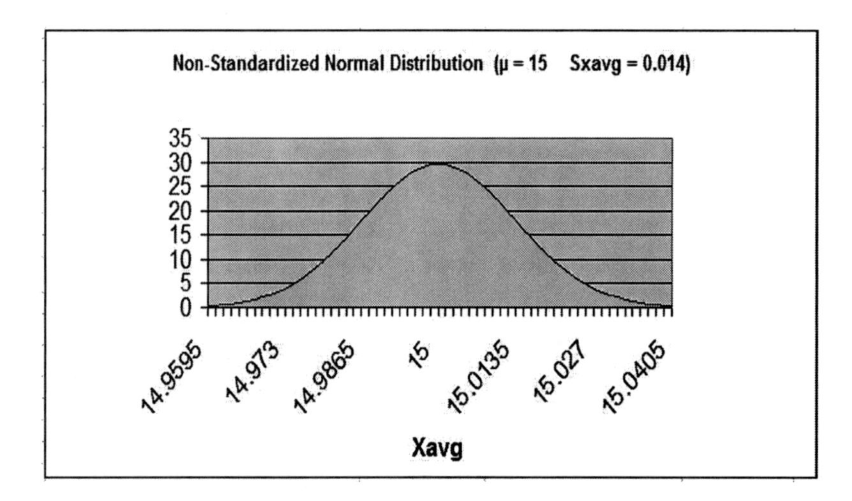

Step 3 - Map the Region of Certainty

The problem requires a 95% Level of Certainty so the Region of Certainty will contain 95% of the area under the Normal curve.

We know that this problem uses a two-tailed test with the Region of Uncertainty split between both outer tails. Each outer tail will contain $\alpha/2$ of the total area, or 0.025 (2.5% of the total area).

The Region of Uncertainty contains 5% of the total area under the Normal curve. The entire **95%** Region of Certainty lies in the center of the Normal curve with 1/2 of the total Region of Uncertainty contained in each outer tail. For a two-tailed test, one tail contains only 1/2 of α. Since α = 0.05 and corresponds to the Level max possibility of error of 5%, each tail contains 0.025 (2.5%) of the total area under the curve.

****

We need to find out how far the boundary of the Region of Certainty is from the Normal curve mean. Calculating the number of standard errors from the Normal curve mean to the outer boundary of the Region of Certainty in either tail for a two-tailed test tail is done as follows:

$Z_{95\%,2\text{-tailed}}$ = NORMSINV(1 - α/2) = NORMSINV(**0.975**) = 1.96

Excel Note - NORMSINV(x) = The number of standard errors from the Normal curve mean to a point right of the Normal curve mean at which x percent of the area under the Normal curve will be to the left of that point. For a two-tailed test, x equals the Level of Certainty plus 1/2 the Level of Uncertainty. This is because the area under the curve to the left of x will contain the entire Region of Certainty and the 1/2 part of the Region of Uncertainty that exists in the outer left tail.

Additional note - For a two-tailed test, NORMSINV(x) can be used to calculate the number of standard errors from the Normal curve mean to the boundary of the Region of

Certainty for each boundary in the left or the right tail. For the two-tailed test, both outer boundaries will be the same distance from the Normal curve mean.

The Region of Certainty extends to the left and to the right of the Normal curve mean of $\mathbf{x_{avg}}$ = 15 by 1.96 standard errors.

One standard error = $\mathbf{s_{xavg}}$ = 0.014, so:

1.96 standard errors = (1.96) * (0.014) = 0.027

The outer boundaries of the Region of Certainty have the values:
$= \mu +/- \mathbf{Z_{95\%,2\text{-tailed}}} * \mathbf{s_{xavg}}$

which equals 15 +/- (1.96) * (0.014) = 15 +/- 0.027 = 14.973 to 15.027

These points (14.973 and 15.027) are 1.96 standard errors from the Normal curve mean of $\mathbf{x_{avg}}$ = 15

These points (14.973 and 15.027) are left and right boundaries of the 95% Region of Certainty on the Normal curve.

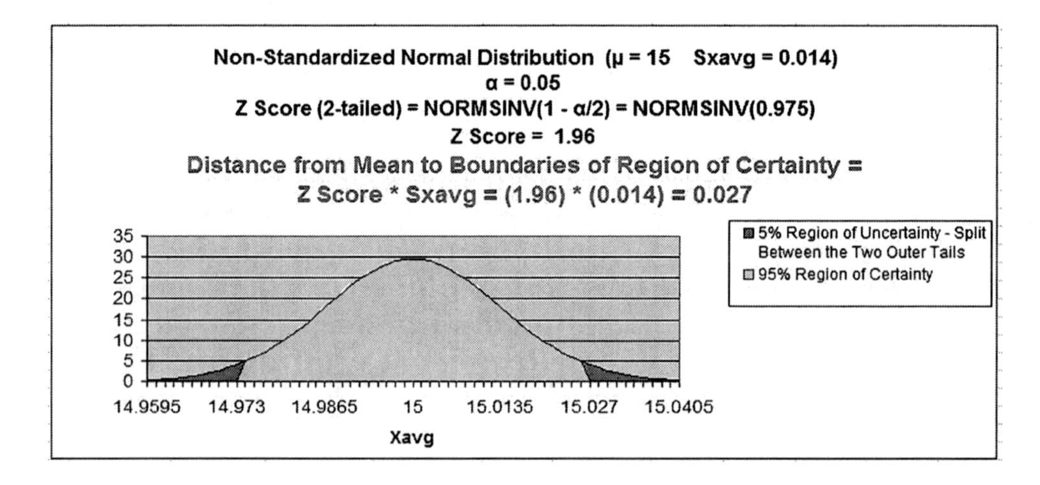

Step 4 - Perform Critical Value and p-Value Tests

a) Critical Value Test

The Critical Value Test is the final test to determine whether to reject or not reject the Null Hypothesis. The p Value Test, described below, is an equivalent alternative to the Critical Value Test.

The Critical Value test tells whether the value of the actual variable, xavg, falls inside or outside of the **Critical Value,** which is the **boundary between the Region of Certainty and the Region of Uncertainty**.

If the actual value of the distributed variable, $\mathbf{x_{avg}}$, falls within the Region of Certainty, the Null Hypothesis is not rejected.

If the actual value of the distributed variable, $\mathbf{x_{avg}}$, falls outside of the Region of Certainty and, therefore, into the Region of Uncertainty, the Null Hypothesis is rejected and the Alternate Hypothesis is accepted.

In this case, the actual value of the variable, $\mathbf{x_{avg}} = 14.982$

The actual value of the variable $\mathbf{x_{avg}} = 14.982$ and is therefore to the right of (inside of) the outer left Critical Value (14.973), which is the boundary between the Regions of Certainty and Uncertainty in the left tail.

The actual value of the variable $\mathbf{x_{avg}}$ is inside the Region of Certainty and therefore inside the Critical Value.

We therefore do not reject the Null Hypothesis, which states that the two means are the same within 5% chance of error. The manufacturer's claim appears to be valid.

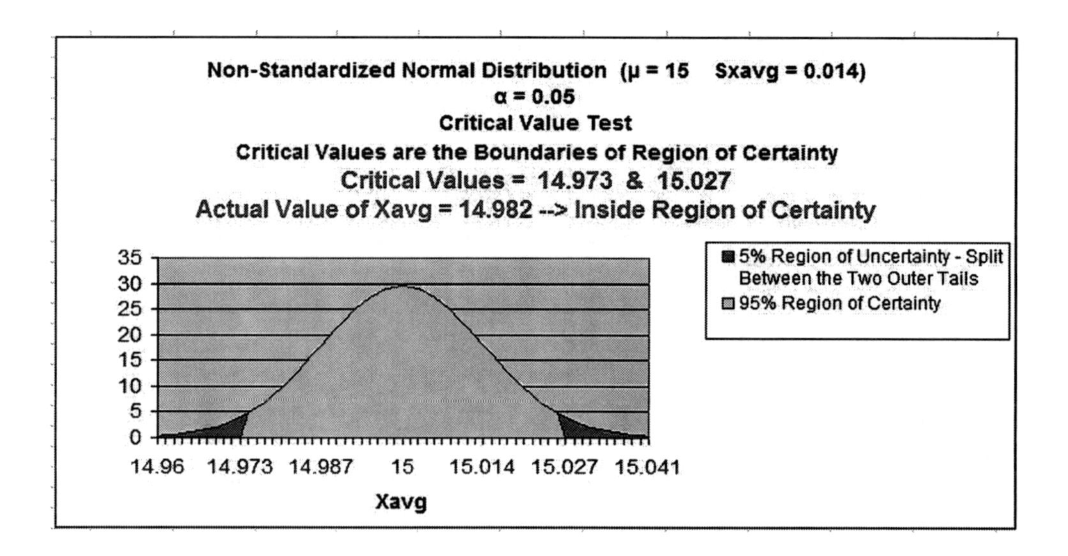

b) p Value Test

The p Value Test is an equivalent alternative to the Critical Value Test and also tells whether to reject or not reject the Null Hypothesis.

The p Value equals the percentage of area under the Normal curve that is in the tail outside of the actual value of the variable $\mathbf{x_{avg}}$.

For a one-tailed test, if the p Value is larger than α, the Null Hypothesis is not rejected. For a two-tailed test, if the p Value is larger than **α**/2, the Null Hypothesis is not rejected.

For a one-tailed test, the Region of Uncertainty is contained entirely in one tail. Therefore the curve area contained by the Region of Uncertainty in that tail equals **α**.

For a two-tailed test, the Region of Uncertainty is split between both tails. Therefore the curve area contained by the Region of Uncertainty in either tail equals **α**/2.

The p Value for the actual value of the distributed variable, which in this case is less than the mean (falls to the left of the mean **in the left tail**), is:

p $Value_{xavg}$ = NORMSDIST([$\mathbf{x_{avg}}$ - $\boldsymbol{\mu}$] / $\mathbf{s_{xavg}}$)

Excel note - NORMSDIST(x) calculates the total area under the Normal curve to the left of the point that is x standard errors to the right of the Normal curve mean.

p $Value_{xavg}$ = NORMSDIST((14.982 - 15) / 0.014) = NORMSDIST(-0.018/0.014) = 0.099

The p Value (0.099) is greater than α/2 (0.025), so the Null Hypothesis is not rejected.

For a two-tailed test---> When the p Value is greater than α/2, the actual value of the distributed variable falls inside the Region of Certainty and the Null Hypothesis is not rejected.

This is the case here.

Following is the p Value graph. The Excel graphing function sometimes produces slightly tilted lines on graphs. The lines should be vertical but are often not.

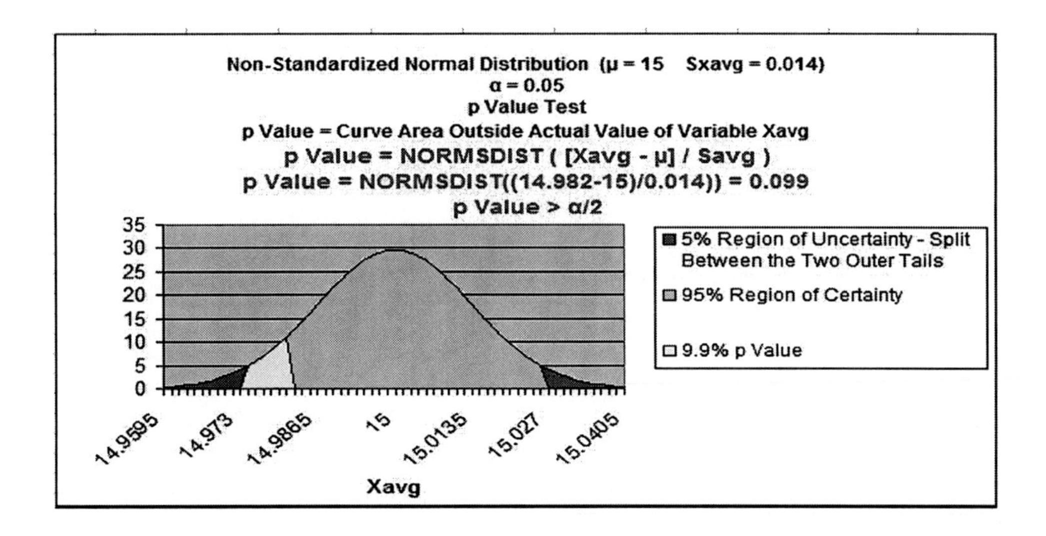

Problem 2 - One-Tailed, One-Sample, Unpaired Hypothesis Test of Mean - Testing whether a delivery time has gotten worse

Problem: A furniture company states that its average delivery time is 15 days with a (population) standard deviation of 4 days. A random sample of 50 deliveries showed an average delivery time of 17 days. Determine within 98% certainty (0.02 significance level) whether delivery time has increased.

We know that this is a **test of mean** and not proportion because **each individual sample taken can have a wide range of values**: Any delivery time sample measurement from 12 to 18 days is probably reasonable.

We know that this is a **one-tailed test** because we are trying to **determine if the "After Data" mean delivery time is larger than the "Before Data" mean delivery time,** not whether the mean delivery times are merely different.

We know that only **one sample** needs to be taken because the population data being tested is already available.

This is **unpaired data** because groups are sampled independently.

"Before Data"	"After Data"
μ = "Before Data" mean = 15	X_{avg} = "After Data" sample average = 17
σ = "Before Data" population standard deviation = 4	
	n = "After Data" Sample size = 50
α = Level of Significance = 0.02 ---> 2% Max chance of error ---> 98% Level of Certainty Required	

This problem can be solved using the standard four-step method for Hypothesis testing.

Step 1 - Create the Null and Alternate Hypotheses

The Null Hypothesis normally states that both means are the same. If the "Before Data" population mean, μ, equals the "After Data" sample mean, $\mathbf{x_{avg}}$, then $\mathbf{x_{avg}} = \boldsymbol{\mu} = 15$

The Null Hypothesis states that both means are the same, which is equivalent to:

The Null Hypothesis, which states that $\mathbf{x_{avg}}$ is the same as **μ** (which is 15), is as follows:

Null Hypothesis, H_o ----> $\mathbf{x_{avg}} = 15$

The Alternate Hypothesis states that the After Data mean is larger, which is equivalent to:

The Alternate Hypothesis, which states that $\mathbf{x_{avg}}$ is larger than **μ** (which is 15), is as follows:

Alternate Hypothesis, H_1 ----> $\mathbf{x_{avg}} > 15$

For this one-tailed test, the Alternative Hypothesis states that the value of the distributed variable $\mathbf{x_{avg}}$ is larger than the value of 15 stated in the Null Hypothesis,

The **Region of Uncertainty** will be entirely in the **right outer tail**.

Note - the Alternative Hypothesis determines whether the Hypothesis test is a one-tailed test or a two-tailed test as follows:

One-tailed test ----> (Value of variable) **is greater than** OR is less than (Constant)

Two-tailed test ----> (Value of variable) **does not equal** (Constant)

Step 2 - Map the Normal Curve

We now create a Normal curve showing a distribution of the same variable that is used by the Null Hypothesis, which is $\mathbf{x_{avg}}$.

The mean of this Normal curve will occur at the same value of the distributed variable as stated in the Null Hypothesis.

Since the Null Hypothesis states that $\mathbf{x_{avg}} = 15$, the Normal curve will map the distribution of the variable $\mathbf{x_{avg}}$ with a mean of $\mathbf{x_{avg}} = 15$.

This Normal curve will have a standard error that is calculated as the standard error of a sample taken from a population is normally calculated, as follows:

Sample Standard Error = $\mathbf{s_{xavg}} = \boldsymbol{\sigma}$ / SQRT($\mathbf{n}$) = 4 / SQRT(50) = 0.566

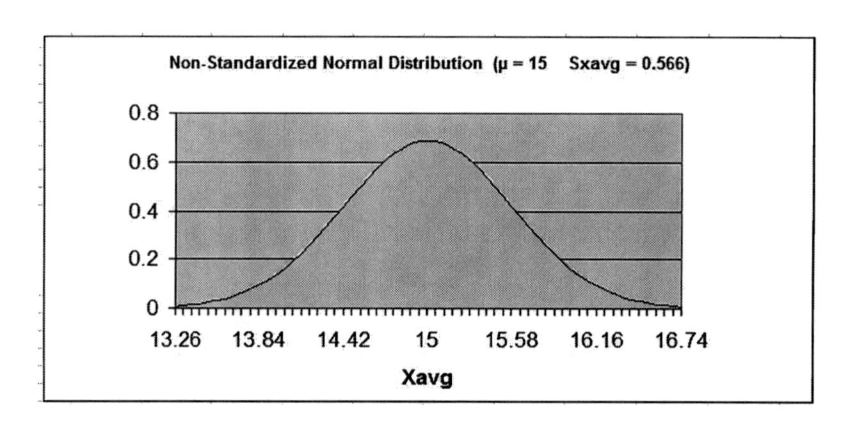

Step 3 - Map the Region of Certainty

The problem requires a 98% Level of Certainty so the Region of Certainty will contain 98% of the area under the Normal curve.

We know that this problem uses a one-tailed test with the Region of Uncertainty entirely contained in the outer right tail.

The Region of Uncertainty contains 2% of the total area under the Normal curve. The entire **98%** Region of Certainty lies to the left of the 2% Region of Uncertainty, which is entirely contained in the outer right tail.

We need to find out how far the boundary of the Region of Certainty is from the Normal curve mean. Calculating the number of standard errors from the Normal curve mean to the outer boundary of the Region of Certainty in the right tail for a one-tailed test is done as follows:

$Z_{98\%,1\text{-tailed}}$ = NORMSINV(1 - α) = NORMSINV(**0.98**) = 2.05

Excel Note - NORMSINV(x) = The number of standard errors from the Normal curve mean to a point right of the Normal curve mean at which x percent of the area under the Normal curve will be to the left of that point.

Additional note - For a one-tailed test, NORMSINV(x) can be used to calculate the number of standard errors from the Normal curve mean to the boundary of the Region of Certainty whether it is in the left or the right tail.

The Region of Certainty extends to the right of the Normal curve mean of $\mathbf{x_{avg}}$ = 15 by 2.05 standard errors.

One standard error = $\mathbf{s_{xavg}}$ = 0.566, so:

2.05 standard errors = (2.05) * (0.566) = 1.16

The outer boundary of the Region of Certainty has the value = $\mu + Z_{98\%,1\text{-tailed}} * s_{xavg}$

which equals 15 + (2.05) * (0.566) = 15 + 1.16 = 16.16

The point, 16.16, is 2.05 standard errors from the Normal curve mean of x_{avg} = 15

This point, 16.16, is the right boundary of the 98% Region of Certainty on the Normal curve.

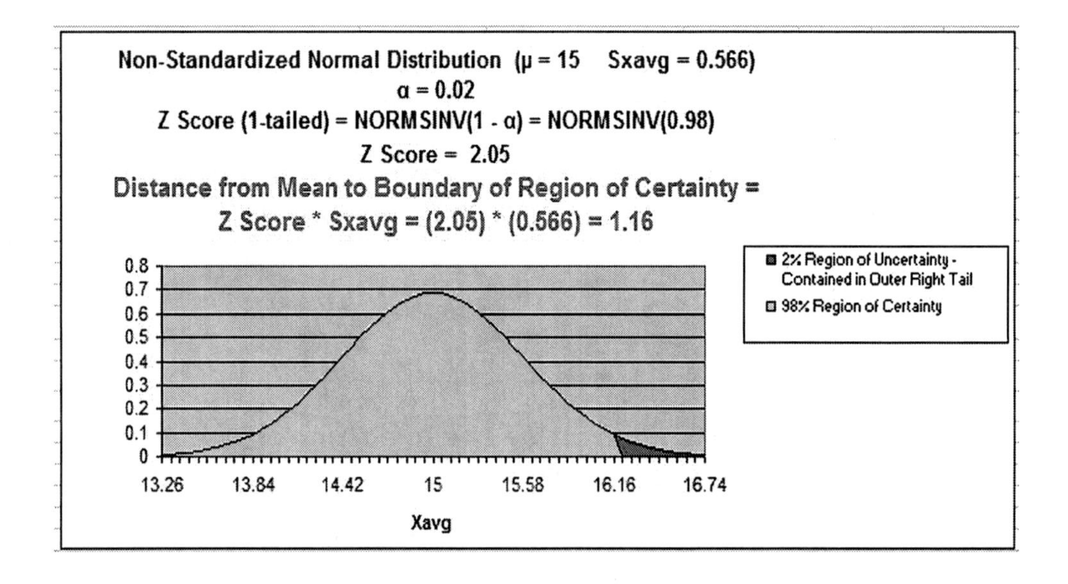

Step 4 - Perform Critical Value and p-Value Tests

a) Critical Value Test

The Critical Value Test is the final test to determine whether to reject or not reject the Null Hypothesis. The p Value Test, described later, is an equivalent alternative to the Critical Value Test.

The Critical Value test tells whether the value of the actual variable, xavg, falls inside or outside of the **Critical Value**, which is the **boundary between the Region of Certainty and the Region of Uncertainty**.

If the actual value of the distributed variable, x_{avg}, falls within the Region of Certainty, the Null Hypothesis is not rejected.

If the actual value of the distributed variable, x_{avg}, falls outside of the Region of Certainty and, therefore, into the Region of Uncertainty, the Null Hypothesis is rejected and the Alternate Hypothesis is accepted.

In this case, the actual value of the variable, x_{avg} = 17

The actual value of the variable x_{avg} = 17 and is therefore to the right of (outside of) the outer right Critical Value (16.16), which is the boundary between the Regions of Certainty and Uncertainty in the right tail.

The actual value of the variable x_{avg} is outside the Region of Certainty and therefore outside the Critical Value.

We therefore reject the Null Hypothesis and accept the Alternate Hypothesis which states that delivery time has increased, with a maximum possible error of 2%. This is shown in the following diagram:

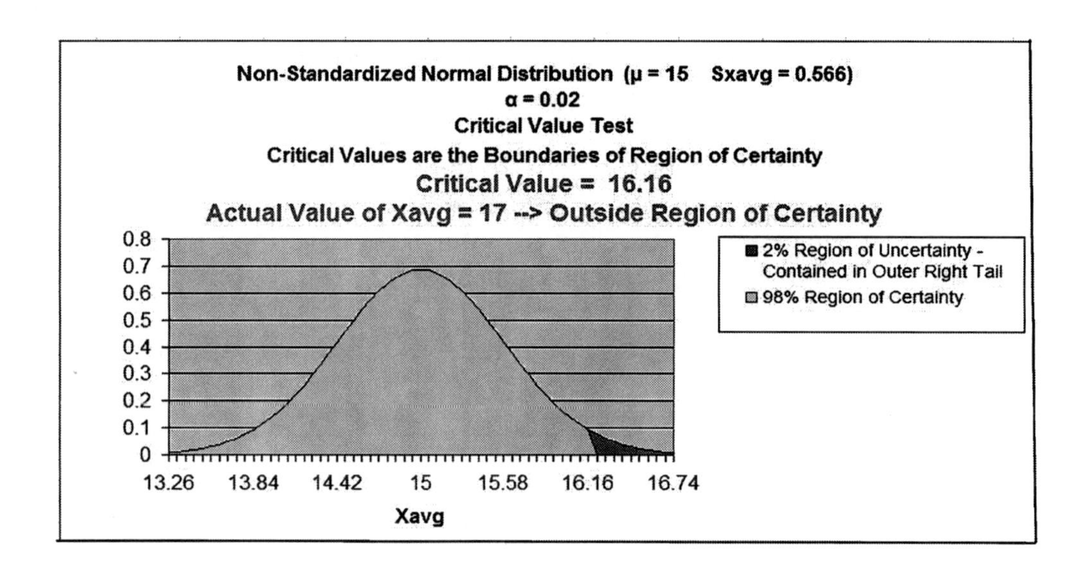

b) p Value Test

The p Value Test is an equivalent alternative to the Critical Value Test and also tells whether to reject or not reject the Null Hypothesis.

The p Value equals the percentage of area under the Normal curve that is in the tail outside of the actual value of the variable $\mathbf{x_{avg}}$.

For a one-tailed test, if the p Value is larger than α, the Null Hypothesis is not rejected. For a two-tailed test, if the p Value is larger than **α**/2, the Null Hypothesis is not rejected.

For a one-tailed test, the Region of Uncertainty is contained entirely in one tail. Therefore the curve area contained by the Region of Uncertainty in that tail equals **α**.

For a two-tailed test, the Region of Uncertainty is split between both tails. Therefore the curve area contained by the Region of Uncertainty in that tail equals **α**/2.

The p Value for the actual value of the distributed variable, which in this case is greater than the mean (falls to the right of the mean **in the right tail**), is:

p $Value_{xavg}$ = **1** - NORMSDIST(**[**x_{avg} - **μ]** / s_{xavg})

Excel note - NORMSDIST(x) calculates the total area under the Normal curve to the left of the point that is x standard errors to the right of the Normal curve mean.

p $Value_{xavg}$ = 1 - NORMSDIST((17 - 15) / 0.566)

= 1 - NORMSDIST(2/0.566)

= 0.0002

The p Value (0.0002) is less than **α** (0.02), so the Null Hypothesis is rejected and the Alternate Hypothesis is accepted..

For a one-tailed test---> When the p Value is less than α, the actual value of the distributed variable falls outside the Region of Certainty and the Null Hypothesis is rejected.

This is the case here as is shown in the following diagram on the next page.

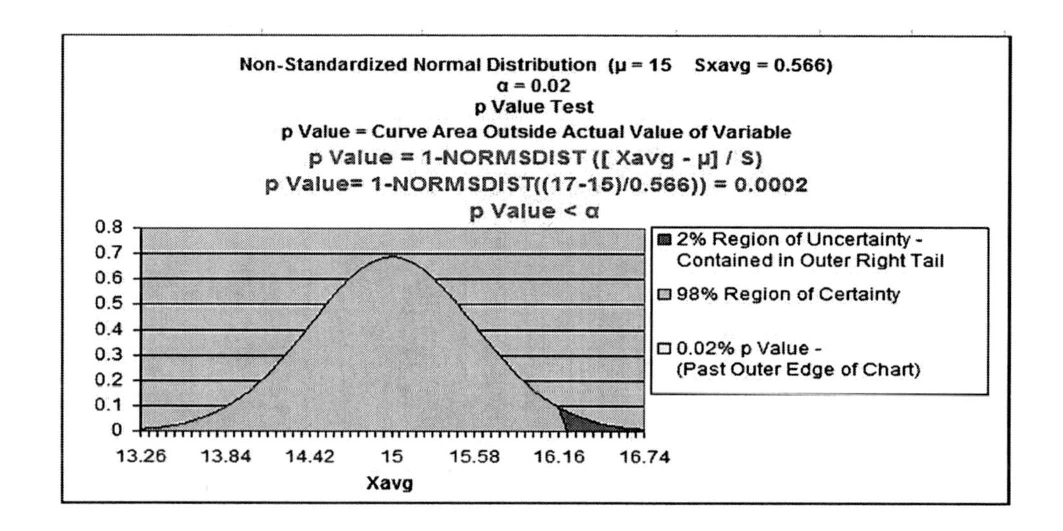
Non-Standardized Normal Distribution (μ = 15 Sxavg = 0.566)
α = 0.02
p Value Test
p Value = Curve Area Outside Actual Value of Variable
p Value = 1-NORMSDIST ([Xavg - μ] / S)
p Value= 1-NORMSDIST((17-15)/0.566)) = 0.0002
p Value < α
0.8
0.7
0.6
0.5
0.4
0.3
0.2
0.1
0
13.26
13.84
14.42
15
15.58
16.16
16.74
Xavg
2% Region of Uncertainty - Contained in Outer Right Tail
98% Region of Certainty
0.02% p Value - (Past Outer Edge of Chart)

Problem 3 - Two-Tailed, Two-Sample, Unpaired Hypothesis Test of Mean - Testing whether wages are the same in two areas

Problem: A survey was taken of wages of unskilled laborers in two areas. The sample data is below. Determine with no more than a 2% chance of error whether or not the mean wages are the same.

We know that this is a **test of mean** and not proportion because **each individual sample taken can have a wide range of values**: Any wage sample measurement from $280 to $320 is probably reasonable.

We know that this is a **two-tailed test** because we are trying to **determine if the mean wages from Area 1 and Area 2 are merely different**, not whether one mean is larger or smaller than the other.

We know that **two samples** need to be taken because no data is initially available.

This is **unpaired data** because the groups are sampled independently.

Area 1	Area 2
X_{1avg} = Sample mean wage 1 = $300.01	X_{2avg} = Sample mean wage 1 = $295.21
S_1 = Sample standard deviation 1 = $4.00	S_2 = Sample standard deviation 2 = $4.50
n_1 = Sample size 1 = 100	n_2 = Sample size 2 = 200
α = Level of significance = 0.02 ---> 2% max chance of error ---> 98% Level of Certainty Required	

This problem can be solved using the standard four-step method for Hypothesis testing.

Step 1 - Create the Null and Alternate Hypotheses

The Null Hypothesis normally states that both populations sampled are the same. If the mean wages from both populations are the same, then $x_{1avg} = x_{2avg}$

The Null Hypothesis states that both mean wages are the same,

which is equivalent to: Null Hypothesis, H_0 ----> $x_{1avg} - x_{2avg} = 0$

****

The Alternate Hypothesis states that the mean wages are different, which is equivalent to:

The Alternate Hypothesis, which states that x1avg is different than x_{2avg}

Alternate Hypothesis, H_1 ----> $x_{1avg} - x_{2avg} \neq 0$

****

For this two-tailed test, the Alternative Hypothesis states that the value of the distributed variable (x1avg - x2avg) does not equal the value of 0 as stated in the Null Hypothesis.

The **Region of Uncertainty** will be split between **both outer tails**.

Note - the Alternative Hypothesis determines whether the Hypothesis test is a one-tailed test or a two-tailed test as follows:

One-tailed test ----> (Value of variable) **is greater than** OR **is less than** (Constant)

Two-tailed test ----> (Value of variable) **does not equal** (Constant)

Step 2 - Map the Normal Curve

We now create a Normal curve showing a distribution of the same variable that is used by the Null Hypothesis, which is (x_{1avg} - x_{2avg}).

The mean of this Normal curve will occur at the same value of the distributed variable as stated in the Null Hypothesis.

Since the Null Hypothesis states that x_{1avg} - x_{2avg} = 0, the Normal curve will map the distribution of the variable (x_{1avg} - x_{2avg}) with a mean of (x_{1avg} - x_{2avg}) = 0

****

The standard error of the difference between two sample variables is approximated as follows:

The approximate standard error of: $s_{(x1avg - x2avg)}$ = SQRT [($s1^2$) / n1 + ($s2^2$) / n2]

Standard Error = $s_{(x1avg - x2avg)}$ = SQRT [(s_1^2) / n1 + (s_2^2) / n2]

Standard Error = $s_{(x1avg - x2avg)}$ = SQRT [(4.00^2) / 100 + (4.50^2) / 200] = 0.51

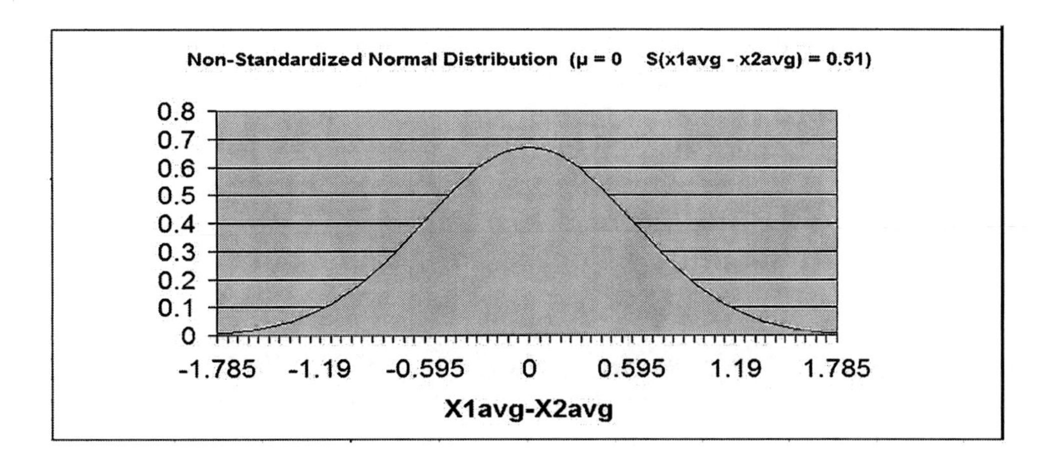

Step 3 - Map the Region of Certainty

The problem requires a 98% Level of Certainty so the Region of Certainty will contain 98% of the area under the Normal curve.

We know that this problem uses a two-tailed test with the Region of Uncertainty split between both outer tails. Each outer tail will contain $\alpha/2$ of the total area, or 0.01 (1.0% of the total area).

The Region of Uncertainty contains 2% of the total area under the Normal curve. The entire **98%** Region of Certainty lies in the center of the Normal curve with 1/2 of the total Region of Uncertainty contained in each outer tail. For a two-tailed test, each tail contains only 1/2 of α. Since α = 0.02 and corresponds to the max possibility of error of 2%, each tail contains 0.01 (1.0%) of the total area under the curve.

We need to find out how far the boundary of the Region of Certainty is from the Normal curve mean. Calculating the number of standard errors from the Normal curve mean to the outer boundary of the Region of Certainty in either tail for a two-tailed test tail is done as follows:

$Z_{98\%,2\text{-tailed}}$ = NORMSINV(1 - α/2) = NORMSINV(**0.99**) = 2.33

Excel Note - NORMSINV(x) = The number of standard errors from the Normal curve mean to a point right of the Normal curve mean at which x percent of the area under the Normal curve will be to the left of that point. For a two-tailed test, x equals the Level of Certainty plus 1/2 the Level of Uncertainty. This occurs because the area under the curve to the left of x will contain the entire Region of Certainty and the 1/2 part of the Region of Uncertainty that exists in the outer left tail.

Additional note - For a two-tailed test, NORMSINV(x) can be used to calculate the number of standard errors from the Normal curve mean to the boundary of the Region of

Certainty for each boundary in the left or the right tail. For the two-tailed test, both outer boundaries will be the same distance from the Normal curve mean.

The Region of Certainty extends to the left and to the right of the Normal curve mean of (x_{1avg} - x_{2avg}) = 0 by 2.33 standard errors.

One standard error = $s_{(x1avg-x2avg)}$ = 0.51, so:

2.33 standard errors = (2.33) * (0.51) = 1.19

The outer boundaries of the Region of Certainty have the values

= μ +/- $Z_{98\%,2\text{-tailed}}$ * $s_{(x1avg-x2avg)}$

which equals 0 +/- (2.33) * (0.51) = +/- 1.19

These points (-1.19 and 1.19) are 2.33 standard errors from the Normal curve mean of (x_{1avg} - x_{2avg}) = 0

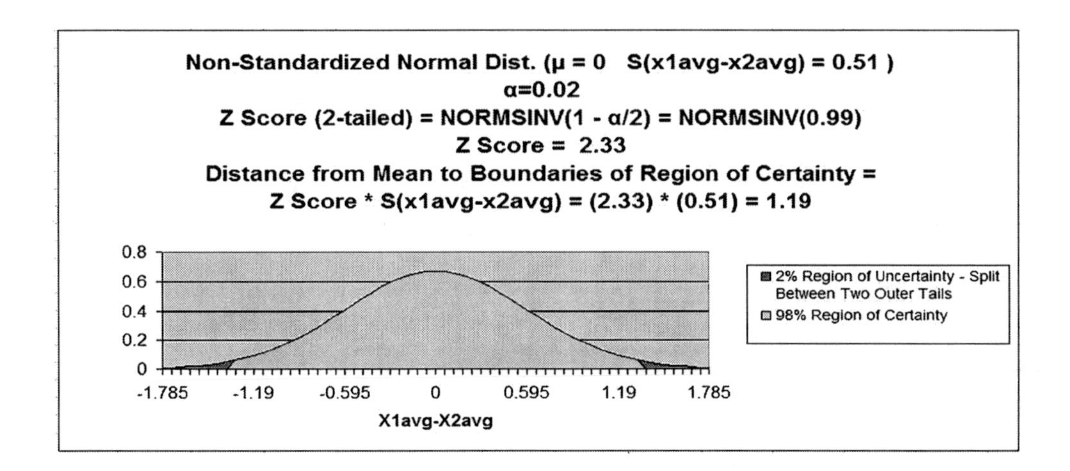

Step 4 - Perform Critical Value and p-Value Tests

a) Critical Value Test

The Critical Value Test is the final test to determine whether to reject or not reject the Null Hypothesis. The p Value Test, described later, is an equivalent alternative to the Critical Value Test.

The Critical Value test tells whether the value of the actual variable, x1avg - x2avg, falls inside or outside of the **Critical Value**, which **is the boundary between the Region of Certainty and the Region of Uncertainty**.

If the actual value of the distributed variable, $\mathbf{x_{1avg}}$ - **x2avg**, falls within the Region of Certainty, the Null Hypothesis is not rejected.

If the actual value of the distributed variable, $\mathbf{x_{1avg}}$ - $\mathbf{x_{2avg}}$, falls outside of the Region of Certainty and, therefore, into the Region of Uncertainty, the Null Hypothesis is rejected and the Alternate Hypothesis is accepted.

In this case, the actual value of the variable, $\mathbf{x_{1avg}}$ - $\mathbf{x_{2avg}}$ =
= \$300.01 - \$295.21 = \$4.80

The actual value of the variable ($\mathbf{x_{1avg}}$ - $\mathbf{x_{2avg}}$) = 4.80 and is therefore to the right of (outside of) the outer right Critical Value (1.19), which is the boundary between the Regions of Certainty and Uncertainty in the right tail.

The actual value of the variable ($\mathbf{x_{1avg}}$ - $\mathbf{x_{2avg}}$) is outside of the Region of Certainty and therefore outside of the Critical Value.

We therefore reject the Null Hypothesis and accept the Alternate Hypothesis, which states that the average wages in the two areas are different with no more than 2% chance of error. This is shown in the following diagram:

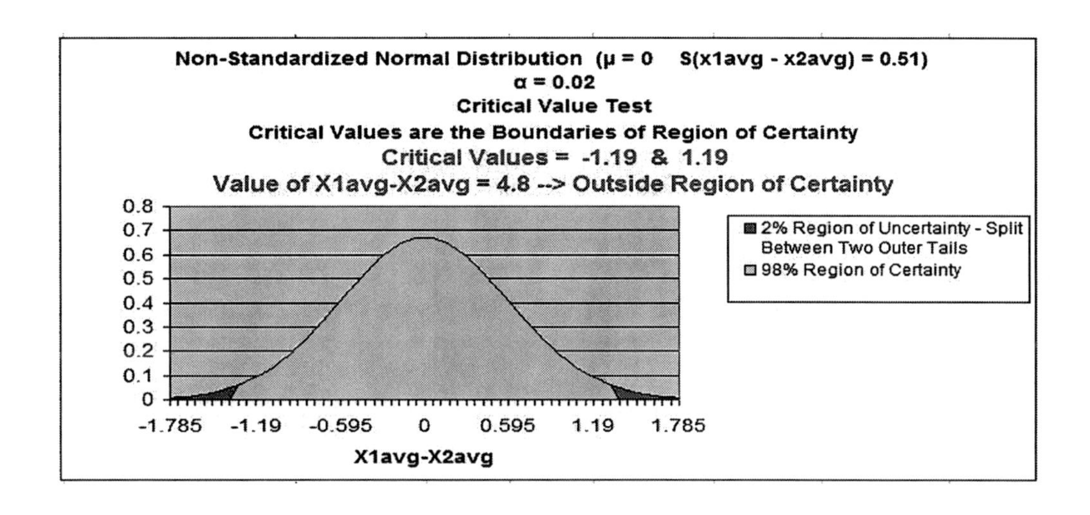

b) p Value Test

The p Value Test is an equivalent alternative to the Critical Value Test and also tells whether or not to reject the Null Hypothesis.

The p Value equals the percentage of area under the Normal curve that is in the tail outside of the actual value of the variable ($\mathbf{x_{1avg}}$ - $\mathbf{x_{2avg}}$).

For a one-tailed test, if the p Value is larger than α, the Null Hypothesis is not rejected.For a two-tailed test, if the p Value is larger than **α**/2, the Null Hypothesis is not rejected.

For a one-tailed test, the Region of Uncertainty is contained entirely in one tail. Therefore the curve area contained by the Region of Uncertainty in that tail equals **α**.

For a two-tailed test, the Region of Uncertainty is split between both tails. Therefore the curve area contained by the Region of Uncertainty in that tail equals **α**/2.

The p Value for the actual value of the distributed variable, which in this case is greater than the mean (falls to the right of the mean **in the right tail**), is:

p $Value_{(x1avg-x2avg)}$ = **1** - NORMSDIST(**[** ($\mathbf{x_{1avg}}$ - $\mathbf{x_{2avg}}$) - **μ]** / $\mathbf{s_{(x1avg - x2avg)}}$)

Excel note - NORMSDIST(x) calculates the total area under the Normal curve to the left of the point that is x standard errors to the right of the Normal curve mean.

p $Value_{(x1avg-x2avg)}$ = 1 - NORMSDIST((4.80 - 0) / 0.51)
= 1 - NORMSDIST(4.80/0.51) ≈ 0

The p Value (0) is less than **α**/2 (0.01), so the Null Hypothesis is rejected and the Alternate Hypothesis is accepted.
For a two-tailed test---> When the p Value is less than α/2, the actual value of the distributed variable falls outside the Region of Certainty and the Null Hypothesis is rejected.

This is the case here.

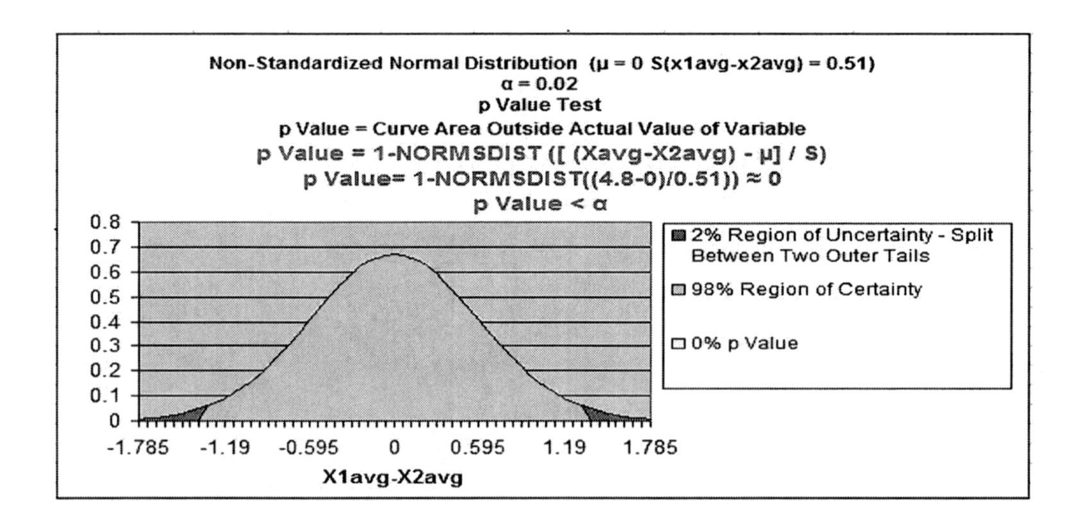

Paired Data

Problem 4 - One-Tailed, Two-Sample, Paired Hypothesis Test of Mean - Testing whether an advertising campaign improved sales

Problem: Determine with 95% certainty whether an advertising campaign increased average daily sales to our large dealer network. Before and After samples of average daily sales were taken with at least 30 dealers.

We know that this is a **test of mean** and not proportion because **each individual sample taken can have a wide range of values**: Any sales sample measurement from 90 to 250 is probably reasonable.

We know that this is a **one-tailed test** because we are trying to **determine if the "After Data" mean sales is larger than the "Before Data" mean sales**, not whether the mean sales are merely different.

We know that **two samples** need to be taken because no data is initially available.

This is **paired data** because each set of "Before" and "After" data came from the same object.

Following on the next page is the data sample:

	BEFORE		AFTER		DIFFERENCE
	Average		Average		DATA
	Daily		Daily		x
DEALER	Sales		Sales		Difference
A	100	-	110	=	10
B	130	-	135	=	5
C	120	-	122	=	2
D	140	-	157	=	17
E	155	-	160	=	5
F	200	-	206	=	6
G	300	-	309	=	9
H	260	-	283	=	23
I	190	-	202	=	12
J	185	-	192	=	7
K	100	-	110	=	10
L	130	-	135	=	5
M	120	-	122	=	2
N	140	-	157	=	17
O	155	-	160	=	5
P	200	-	206	=	6
Q	300	-	309	=	9
R	260	-	283	=	23
S	190	-	202	=	12
T	185	-	192	=	7
U	100	-	110	=	10
V	130	-	135	=	5
W	120	-	122	=	2
X	140	-	157	=	17
Y	155	-	160	=	5
Z	200	-	206	=	6
A1	300	-	309	=	9
B1	260	-	283	=	23
C1	190	-	202	=	12
D1	185	-	192	=	7

Paired data tests involve taking Before and After samples from the same large number (n >30) of objects and performing a Hypothesis test on the differences between the Before and After samples.

In this case, the yellow-highlighted column represents the difference between the Before and After sample of each data pair. The Hypothesis test will be performed on that column of data.

Before we can begin the Hypothesis test, we need to calculate the following parameters of variable x:

Sample size - Use Excel function COUNT
Sample mean - Use Excel function AVERAGE
Sample standard deviation - Use Excel function STDEV

Sample standard error = (Sample standard deviation) / SQRT (Sample size)

"Difference Data"

X_{avg} = "Difference Data" sample average = 9.60
S_{xavg} = Sample standard error = 1.11

n = "Difference Data" Sample size = 30

α = Level of Significance = 0.05 ---> 5% Max chance of error
---> 95% Level of Certainty Required

This problem can be solved using the standard four-step method for Hypothesis testing.

Step 1 - Create the Null and Alternate Hypotheses

The Null Hypothesis normally states that both populations sampled are the same. If the mean sales from both the Before and After Data are the same, then their average difference = 0

The Null Hypothesis states that both Before and After mean sales are the same, which is equivalent to:

Null Hypothesis, H_0 ----> $\mathbf{x_{avg}} = 0$

****

The Alternate Hypothesis states that the After Data mean sales is larger, which is equivalent to:

The Alternate Hypothesis, which states that their average difference, $\mathbf{x_{avg}}$, is larger than 0, is as follows:

Alternate Hypothesis, H_1 ----> $\mathbf{x_{avg}} > 0$

****

For this one-tailed test, the Alternative Hypothesis states that the value of the distributed variable $\mathbf{x_{avg}}$ is larger than the value of 0 stated in the Null Hypothesis,

The **Region of Uncertainty** will be entirely in the **right outer tail**.

Note - the Alternative Hypothesis determines whether the Hypothesis test is a one-tailed test or a two-tailed test as follows:

One-tailed test ----> (Value of variable) **is greater than** OR **is less than** (Constant)

Two-tailed test ----> (Value of variable) **does not equal** (Constant)

Step 2 - Map the Normal Curve

We now create a Normal curve showing a distribution of the same variable that is used by the Null Hypothesis, which is x_{avg}.

The mean of this Normal curve will occur at the same value of the distributed variable as stated in the Null Hypothesis.

Since the Null Hypothesis states that $x_{avg} = 0$, the Normal curve will map the distribution of the variable x_{avg} with a mean of $x_{avg} = 0$.

****

This Normal curve will have a standard error as just calculated as follows:

Standard Error = $s_{xavg} = 1.11$

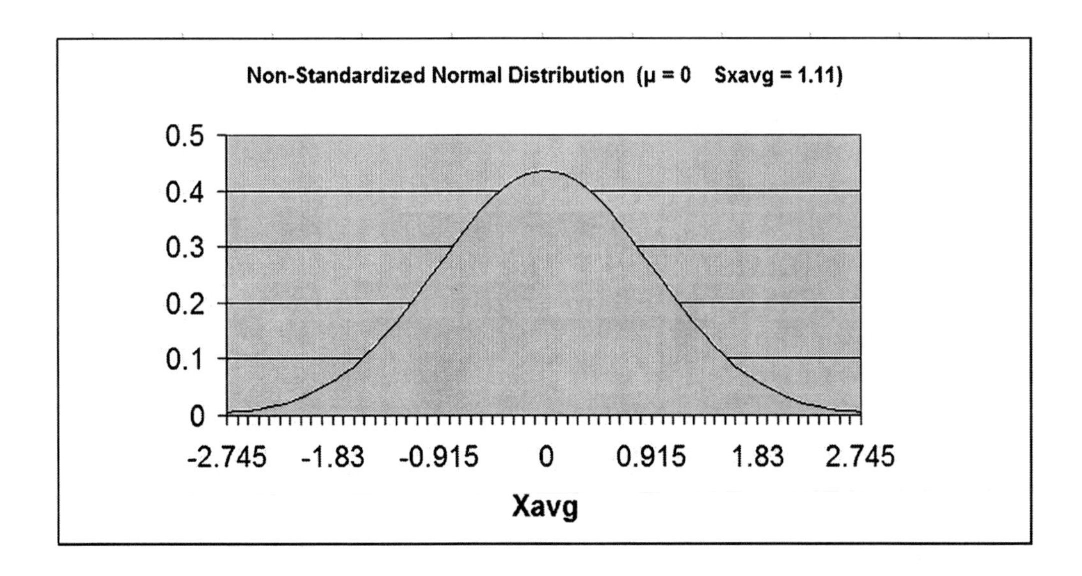

Step 3 - Map the Region of Certainty

The problem requires a 95% Level of Certainty so the Region of Certainty will contain 95% of the area under the Normal curve.

We know that this problem uses a one-tailed test with the Region of Uncertainty entirely contained in the outer right tail.

The Region of Uncertainty contains 5% of the total area under the Normal curve. The entire **95%** Region of Certainty lies to the left of the 5% Region of Uncertainty, which is entirely contained in the outer right tail.

****

We need to find out how far the boundary of the Region of Certainty is from the Normal curve mean. Calculating the number of standard errors from the Normal curve mean to the outer boundary of the Region of Certainty in the right tail for a one-tailed test tail is done as follows:

$Z_{95\%,1\text{-tailed}}$ = NORMSINV(1 - α) = NORMSINV(**0.95**) = 1.65

Excel Note - NORMSINV(x) = The number of standard errors from the Normal curve mean to a point right of the Normal curve mean at which x percent of the area under the Normal curve will be to the left of that point.

Additional note - For a one-tailed test, NORMSINV(x) can be used to calculate the number of standard errors from the Normal curve mean to the boundary of the Region of Certainty whether it is in the left or the right tail.

The Region of Certainty extends to the right of the Normal curve mean of x_{avg} = 0 by 1.65 standard errors.

One standard error = s_{xavg} = 1.11, so:

1.65 standard errors = (1.65) * (1.11) = 1.83

The outer boundary of the Region of Certainty has the value = $\mu + Z_{95\%,\text{one-tailed}} * s_{xavg}$

which equals 0 + (1.65) * (1.11) = 0 + 1.83 = 1.83

The point, 1.83, is 1.65 standard errors from the Normal curve mean of x_{avg} = 0

This point, 1.83, is the right boundary of the 95% Region of Certainty on the Normal curve.

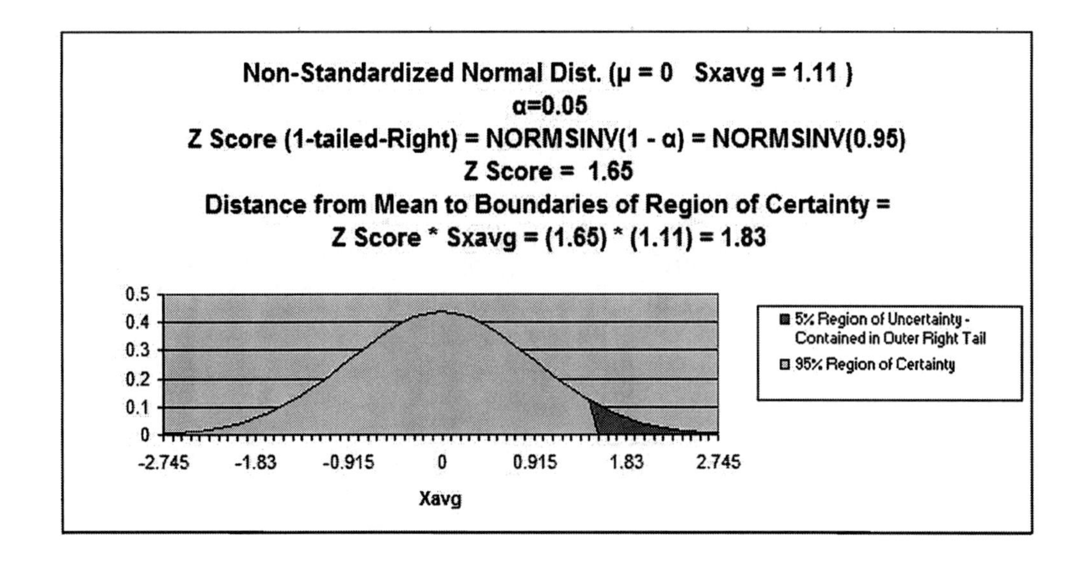

Step 4 - Perform Critical Value and p-Value Tests

a) Critical Value Test

The Critical Value Test is the final test to determine whether to reject or not reject the Null Hypothesis. The p Value Test, described next, is an equivalent alternative to the Critical Value Test.

The Critical Value test tells whether the value of the actual variable, **xavg**, falls inside or outside of the **Critical Value**, which is the **boundary between the Region of Certainty and the Region of Uncertainty**.

If the actual value of the distributed variable, x_{avg}, falls within the Region of Certainty, the Null Hypothesis is not rejected.

If the actual value of the distributed variable, x_{avg}, falls outside of the Region of Certainty and, therefore, into the Region of Uncertainty, the Null Hypothesis is rejected and the Alternate Hypothesis is accepted.

The actual value of the variable x_{avg} = 9.60 and is therefore to the right of (outside of) the outer right Critical Value (1.83), which is the boundary between the Regions of Certainty and Uncertainty in the right tail.

The actual value of the variable x_{avg} is outside the Region of Certainty and therefore outside the Critical Value.

We therefore reject the Null Hypothesis and accept the Alternate Hypothesis which states that average dealer sales have increased after the ad campaign, with a maximum possible error of 5%. This is shown in the following diagram on the next page:

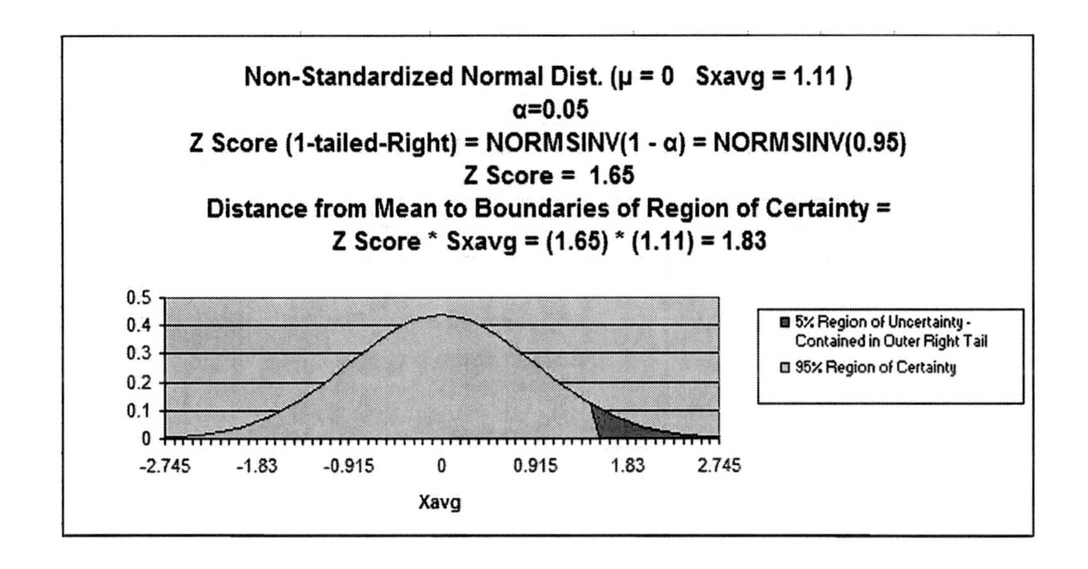

b) p Value Test

The p Value Test is an equivalent alternative to the Critical Value Test and also tells whether to reject or not reject the Null Hypothesis.

The p Value equals the percentage of area under the Normal curve that is in the tail outside of the actual value of the variable $\mathbf{x_{avg}}$.

For a one-tailed test, if the p Value is larger than α, the Null Hypothesis is not rejected. For a two-tailed test, if the p Value is larger than $\boldsymbol{\alpha}/2$, the Null Hypothesis is not rejected.

For a one-tailed test, the Region of Uncertainty is contained entirely in one tail. Therefore the curve area contained by the Region of Uncertainty in that tail equals $\boldsymbol{\alpha}$.

For a two-tailed test, the Region of Uncertainty is split between both tails. Therefore the curve area contained by the Region of Uncertainty in that tail equals $\boldsymbol{\alpha}/2$.

The p Value for the actual value of the distributed variable, which in this case is greater than the mean (falls to the right of the mean **in the right tail**), is:

p $Value_{xavg}$ = 1 - NORMSDIST([x_{avg} - μ] / s_{xavg})

Excel note - NORMSDIST(x) calculates the total area under the Normal curve to the left of the point that is x standard errors to the right of the Normal curve mean.

p $Value_{xavg}$ = 1 - NORMSDIST((9.60 - 0) / 1.11) = 1 - NORMSDIST(9.60/1.11) ≈ 0

The p Value (0) is less than α (0.05), so the Null Hypothesis is rejected and the Alternate Hypothesis is accepted..

For a one-tailed test---> When the p Value is less than α, the actual value of the distributed variable falls outside the Region of Certainty and the Null Hypothesis is rejected.

This is the case here and is shown in the following diagram.

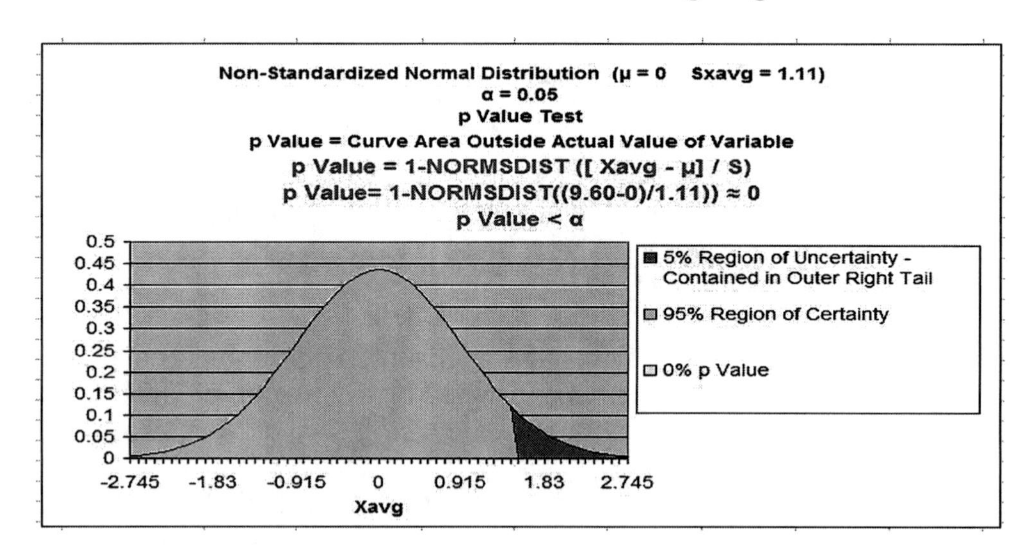

Chapter 9 - Hypothesis Testing of Proportions

Basic Explanation of Hypothesis Testing

Hypothesis testing of proportions is a useful managerial statistical tool to determine whether a change in a population proportion has occurred. For example, this can be used to determine whether the proportion of a population who prefer one election candidate has changed.

Hypothesis testing for change is definitely one of the most useful statistical tools for the business manager. Quite often we need to determine within a small possibility of error whether something measurable has changed. For example, you have implemented a new advertising campaign and you want to determine if sales really have improved. Hypothesis testing can be used in a wide variety of situations where you have Before and After data and you want to determine if real change has occurred. Hypothesis testing is also a great tool to statistically verify whether two groups have the same mean or proportion of something.

Hypothesis testing involves creating two separate Hypotheses and then testing to see which one applies. The two hypotheses are the Null Hypothesis and the Alternate Hypothesis. The Null Hypothesis, often referred to as H_0, usually states that both proportions are the same or that no change has occurred. The Alternate Hypothesis, H_1, states the proportion has changed. In other words, the new proportion is statistically different from the old proportion.

Hypothesis testing uses sample data to verify either the Null or Alternate Hypothesis about a population. The most important goal of Hypothesis testing is to verify the correct hypothesis about a population within a specified degree of certainty. For example, you are conducting a poll to predict the results of an election. You might want to be at least 95% certain whether a candidate's ad campaign had an effect on the proportion of voters who expect to vote for that candidate.

The Four-Step Method to Solving ALL Hypothesis Test Problems

There are many types of Hypothesis tests but fortunately they can all be solved in the same general way using this four-step method:

Step 1 - Create the Null Hypothesis and the Alternate Hypothesis

Step 2 - Map the Normal Curve

Step 3 - Map the Region of Certainty

Step 4 - Perform the Critical Value Test or the p-Value Test

We will cover these steps in more detail later but ALL Hypothesis testing can be done with this four-step method. All of the different Hypothesis test problems presented in this course are solved with the four-step method.

The Four Ways of Classifying ALL Hypothesis Test Problems

1) Mean Testing vs. Proportion Testing

The basic objective of Hypothesis testing is to determine whether the mean or proportion within one group is statistically the same as the mean or proportion within another group. What determines whether a mean is being tested or a proportion is being tested is the number of possible outcomes of each sample taken.

Proportion test samples have only two possible outcomes. For example, if you are comparing the proportion of Republicans in two different cities, each sample has only two possible values; the person sampled either is a Republican or is not.

Mean test samples have multiple possible outcomes. For example, if you are comparing the mean age of people in two different cities, each sample can have numerous values; the person sampled could be anywhere from 1 to 110 years old.

Hypothesis tests of means are computed the same as Hypothesis tests of proportion in every way except one: the calculation of the Sample Standard Error. This difference is significant enough that these two types of Hypothesis tests are analyzed in separate modules of this course.

2) One-Tailed vs. Two-Tailed Testing

All Hypothesis tests are either one-tailed or two-tailed. The number of tails depends on whether the Hypothesis test can determine if the mean or proportion in one sampled group is merely different than in another group or whether it is different in one direction (is either larger or is smaller) than in another group

A one-tailed test is used to determine whether a mean or proportion is different in one direction than another mean or proportion, not that it is merely different.

A one-tailed test would be used to determine if the proportion of Republicans in one city is larger than the proportion of Republicans in another city. Another one-tailed test would be used to determine if the mean age of people in one city is less than the mean age of people in another city.

Two-tailed tests are used to determine whether a mean or proportion is merely different than another mean or proportion. The direction of the difference is not a factor in a two-tailed test, as it is in a one-tailed test.

3) One-Sample vs. Two Sample Testing

Hypothesis testing fundamentally involves comparing one mean or proportion with another mean or proportion. Whether you need to take one sample or two samples depends upon whether you already have original or "Before" data available.

One-sample testing is performed if original or "Before" comparison data is already in place at the start of the test. A one-sample hypothesis test is normally performed to determine whether the original or "Before" data is still valid or whether something has changed.

Two-sample testing is performed if no "Before" data is available or if a comparison is being made but no data is initially available on either side.

To summarize - if data is available for one of the means or proportions being compared, then only one sample is needed for data collection of the other mean or proportion being compared. If no data is available, then two samples must be taken - one for each mean or proportion being compared.

4) Unpaired Data Testing vs. Paired Data Testing

Most Hypothesis testing uses unpaired data testing. Whether data is paired or unpaired depends on whether both samples were collected from the same objects or not.

For data to be Paired data, both samples must have been collected from exactly the same objects. An example of this would be "Before" and "After" sales data taken from the same individual dealers to test whether an ad campaign had increased sales. The dealers are randomly chosen and each dealer provides a single set of "Before" and "After" data. Hypothesis testing is performed on the set of differences between "Before" and "After" numbers for each data pair. Mean testing, but not proportion testing, can be performed using paired data.

Unpaired data samples are group samples collected independently of each other. Groups of unpaired data are treated independently of each other. Separate means and standard errors are calculated from each group. Hypothesis testing is then performed to compare the means or proportions of the two separate groups. The majority of Hypothesis testing is performed using Unpaired data.

Detailed Description of the Four-Step Method for Solving Proportion Testing Problems

The four-step method is used to solve all Hypothesis testing problems. This course module will discuss solving tests of proportion. The previous module of this course covered solving tests of Mean. Both types of tests are solved in the same general way. The main difference between solving for mean and proportion is the calculation of sample standard error. Below is a detailed description of each of the four steps to solving a Hypothesis test of proportion:

Initial Steps

Before solving a hypothesis test problem, you must classify the problem type and lay out the information given in the problem properly.

Problem Classification

Problem Classification: Select the proper choice of each of the four ways that a Hypothesis problem is classified as follows:

1) Mean Testing vs. Proportion Testing

• Proportion test samples have only two possible outcomes.

• Mean test samples have multiple possible outcomes.

2) One-Tailed vs. Two-Tailed Testing

• Two-tailed tests determine whether two means are merely different.

• One-tailed tests determine whether one mean is different in one direction.

3) One-Sample vs. Two-Sample Testing

• One sample is taken if original or "Before" comparison data is available.

• Two samples are taken if no comparison data is available.

4) Unpaired Data Testing vs. Paired Data Testing

- Paired data testing can be performed if "Before" and "After" data is collected from the same objects. Mean testing can be performed on paired data. Proportion testing cannot.

- Unpaired data testing is performed on data collected in groups.

- Paired data testing is performed using only mean testing, not proportion testing.

Information Layout

Information Layout: Listing the given information in a problem properly greatly expedites problem solving. Here is a list of given information that needs to be laid out before solving:

1) Level of Significance

The Level of Significance, α, is equivalent to the maximum possibility of error. The Level of Significance can also be derived from the Required Level of Certainty.

2) Existing Comparison Data

This includes an existing population that is being verified or the "Before" mean that will be compared to the "After" mean. The existing mean data will normally also contain standard deviation or standard error information.

3) Comparison Sample Data

Any sample data will include the sample mean and sample size. Sample proportion data will not include sample standard deviation or sample standard error because individual samples in proportion testing are binary (they can have only two possible values). Standard error is, however, calculated when the Normal curve is mapped (Step 2 of the 4-step process discussed shortly)

The Four Steps to Hypothesis Testing

Descriptions of each of the four steps are here but if you would like to see these steps being applied to all of the different types of proportion Hypothesis tests, look further to see all problems completed with these four steps. Mean Hypothesis testing using these four steps is shown and explained in the preceding module of this course. Directly below is a description of Proportion Hypothesis testing.

Step 1 - Create the Null and Alternate Hypotheses

The Null Hypothesis states that both means are the same.

For a two-sample test, the proportions of both samples, p1avg and p2avg, are being compared. The Null Hypothesis states that they are both equal as follows:

Null Hypothesis, H_o -----> $p_{1avg} - p_{2avg} = 0$

For a one-sample test, the proportion of the sample taken, p_{avg}, **is compared to the Constant** that the original or "Before" proportion, μ_p, was measured to be. The Null Hypothesis states that both are equivalent as follows:

Null Hypothesis, H_o ----> p_{avg} = **Constant**

The Alternate Hypothesis states that both means are different.

A **two-tailed test** states that the two means are merely different, as follows:

- One-sample, two-tailed test Alternate hypothesis

H_1 ----> $\mathbf{p_{avg}} \neq$ **Constant**

- Two-sample, two-tailed test Alternate hypothesis

H_1 ----> $\mathbf{p_{1avg}} - \mathbf{p_{2avg}} \neq$ **Constant**

A **one-tailed test** states that the two means are different in one direction as follows:

- One-sample, one-tailed test Alternate hypothesis

H_1 ----> $\mathbf{p_{avg}} >$ **Constant OR** $\mathbf{p_{avg}} <$ **Constant** ....

- Two-sample, one-tailed test Alternate hypothesis

H_1 ----> $\mathbf{p_{1avg}} - \mathbf{p_{2avg}} >$ **Constant OR** $\mathbf{p_{1avg}} - \mathbf{p_{2avg}} <$ **Constant**

To summarize:

One-tailed test ----> (Value of variable) **is greater than** OR **is less than** (Constant)

Two-tailed test ----> (Value of variable) **does not equal** (Constant)

Step 2 - Map the Normal Curve

We now create a Normal curve showing a distribution of the same variable that is used by the Null Hypothesis, which is p_{avg} or (p_{1avg} - p_{2avg})

The mean of this Normal curve will occur at the same value of the distributed variable as stated in the Null Hypothesis.
with a mean = $\mu_{(pavg)}$ = **Constant**

OR

with a mean = $\mu_{(p1avg-p2avg)}$ = **Constant**

This Normal curve will have a standard error whose calculation depends on whether the mean test uses one sample or two samples as follows:

Please note that the method of calculating the Standard Error is the major difference between proportion tests and mean tests.

One Sample Test

Sample Standard Error = s_{pavg} = SQRT(p_{avg} * q_{avg} / **n**)

$q_{avg} = 1 - p_{avg}$

n = sample size

Two Sample Test

A Normal curve for the distribution of the variable (p_{1avg} - p_{2avg})

which has a mean = $\mu_{(p1avg-p2avg)}$ = **Constant**

and a standard error = $s_{p1avg-p2avg}$ = SQRT [$p_{weighted}$ * $q_{weighted}$ (1 / n1 + 1 / n2)]

with:

$q_{weighted}$ = (n1*p_{1avg} + n2*p_{2avg}) / (n1 + n2)

$q_{weighted}$ = 1 - $p_{weighted}$

n1 and n2 are sample sizes

Step 3 - Map the Region of Certainty

The **Region of Certainty** is the percentage of area under the Normal curve that corresponds with the degree of certainty required by the problem. For example, if the problem requires at least 95% certainty, the Region of Certainty will contain 95% of the area under the Normal curve.

The remainder of the area will be contained in the Region of Uncertainty. The **Region of Uncertainty** is the percentage of area under the Normal curve that corresponds with the maximum allowed possibility of error. For example, if the problem requires at least 95% certainty, then the max chance of error is 5%. The Region of Uncertainty would therefore contain 5% of the total area under the Normal curve.

The area in the Region of Uncertainty corresponds to **α** (alpha). For example, if the problem allows a max chance of error of 5%, then 5% of the total area will be contained in the Region of Uncertainty and α = 0.05.

For a Two-Tailed Test:

The **Region of Uncertainty** will be split between **both outer tails**. Each outer tail will contain $\alpha/2$ of the total area under the Normal curve.

For a One-Tailed Test:

The **Region of Uncertainty** will be contained in **one outer tail**. That outer tail will contain α of the total area under the Normal curve.

The Alternate Hypothesis determines whether the Region of Uncertainty is contained in the left or the right outer tail.

The **Region of Uncertainty** will be contained in the **right outer tail** If: Alternate Hypothesis states ---> (Value of variable) **is greater than** (Constant)

The **Region of Uncertainty** will be contained in the **left outer tail** If: Alternate Hypothesis states ---> (Value of variable **is less than** (Constant)

Mapping the Region of Certainty for a Two-Tailed Test

Mapping the Region of Certainty means calculating how far one or both boundaries of the Region of Certainty are from the Normal curve' s mean in its center. The Region of Certainty has only one outer boundary in a one-tailed test but has two outer boundaries in a two-tailed test.

The Region of Certainty for a two-tailed test is located in the middle of the Normal curve in between the equal Regions of Uncertainty in each outer tail. The left and right outer boundaries of the Region of Certainty are both the same distance from the Normal curve mean. The number of standard errors that either of these two outer boundaries is from the mean is calculated as follows in this example: For a two-tailed test with a required level of certainty of **95%**, the number of standard errors from the Normal curve mean to either of the outer boundaries of the Region of Certainty is calculated in Excel as follows:

$z_{95\%,2\text{-tailed}}$ = NORMSINV(1 - α/2) = NORMSINV(**0.975**) = 1.96

Excel Note - NORMSINV(x) = The number of standard errors from the Normal curve mean to a point right of the Normal curve mean at which x percent of the area under the Normal curve will be to the left of that point. For a two-tailed test, x equals the Level of Certainty plus 1/2 the Level of Uncertainty. This occurs because the area under the curve to the left of x will contain the entire Region of Certainty and the 1/2 part of the Region of Uncertainty that exists in the outer left tail.

The Region of Certainty extends to the left and to the right of the Normal curve mean by 1.96 standard errors.

If, for example, one standard error = s_{pavg} = 0.51, then:

1.96 standard errors = (1.96) * (0.51) = 0.9996

The outer boundaries of the Region of Certainty have the values =
= μ +/- $Z_{95\%,2\text{-tailed}}$ * s_{pavg}

If the Normal curve mean, which is established in the Null Hypothesis equals 0, then the outer boundaries of the Region of Certainty have the values 0 +/- (1.96) * (0.51) = +/- 0.9996

These points (-0.9996 and 0.9996) are 1.96 standard errors from the Normal curve mean of p_{avg} = 0

Mapping the Region of Certainty for a One-Tailed Test

The Region of Certainty for a one-tailed test has only one outer boundary. The entire Region of Uncertainty for the one-tailed test is contained in only one outer tail. The remainder of the area under the Normal curve makes up the Region of Certainty.

The Excel function for calculating the distance from the Normal curve mean to the one boundary of the Region of Certainty is shown in this example:

For a one-tailed test with a required level of certainty of **95%**, the number of standard errors from the Normal curve mean to the outer boundary of the region of Certainty is calculated in Excel as follows:

$Z_{95\%,1\text{-tailed}}$ = NORMSINV(1 - α) = NORMSINV(**0.95**) = 1.65

Excel Note - NORMSINV(x) = The number of standard errors from the Normal curve mean to a point right of the Normal curve mean at which x percent of the area under the Normal curve will be to the left of that point. For a one-tailed test, x equals the Level of Certainty.

Important Excel Note ---> The above Excel function calculates the number of standard errors from the mean to the boundary of the Region of Certainty regardless of whether that boundary is on the left side or the right side of the mean.

If the **Region of Uncertainty** is contained in the **right outer tail,** then the boundary of the Region of Certainty is also in the right tail and therefore has a greater value than the Normal curve mean, μ. The value of boundary for the 95% Region of Certainty in the right tail for a one-tailed test is:

$\mu + Z_{95\%,1\text{-tailed}} * s_{pavg}$

The 95% Region of Certainty for a one-tailed test extends to the right of the Normal curve mean by 1.65 standard errors.

If, for example, one standard error = s_{pavg} = 0.51, then:

1.65 standard errors = (1.65) * (0.51) = 0.8415

The one and only outer boundary of this 95% Region of Certainty has the value = $\mu + \mathbf{Z_{95\%,1\text{-}tailed}} * \mathbf{s_{pavg}}$

If the Normal curve mean, which is established in the Null Hypothesis equals 0, then the outer boundaries of this 95% Region of Certainty have the values 0 + (1.65) * (0.51) = 0.8415

This point (0.8415) is 1.65 standard errors to the right of the Normal curve mean of $\mathbf{p_{avg}}$ = 0

If the **Region of Uncertainty** is contained in the **left outer tail**: then the boundary of the Region of Certainty is also in the left tail and therefore has a lower value than the Normal curve mean, μ. The value of the boundary of the Region of Certainty in the left tail for a one-tailed test is:

$\mu - \mathbf{Z_{95\%,1\text{-}tailed}} * \mathbf{s_{pavg}}$

The Region of Certainty extends to the left of the Normal curve mean by 1.65 standard errors.

If, for example, one standard error = $\mathbf{s_{pavg}}$ = 0.51, then:

1.65 standard errors = (1.65) * (0.51) = 0.8415

The one and only outer boundary of this Region of Certainty has the value = $\mu - \mathbf{Z_{95\%,1\text{-}tailed}} * \mathbf{s_{pavg}}$

If the Normal curve mean, which is established in the Null Hypothesis, equals 0, then the outer boundary of the Region of Certainty has the value: 0 - (1.65) * (0.51) = -0.8415

This point (-0.8415) is 1.65 standard errors to the left of the Normal curve mean of $\mathbf{p_{avg}}$ = 0

Step 4 - Perform Critical Value and p-Value Tests

a) Critical Value Test

The Critical Value Test is the final test to determine whether to reject or not reject the Null Hypothesis. The p Value Test, described below, is an equivalent alternative to the Critical Value Test.

The Critical Value test tells whether the value of the actual variable, pavg, falls inside or outside of the **Critical Value**, which is the **boundary between the Region of Certainty and the Region of Uncertainty**.

If the actual value of the distributed variable, (p_{avg}) or (p_{1avg} - p_{2avg}), falls within the Region of Certainty, the Null Hypothesis is not rejected.

If the actual value of the distributed variable, (p_{avg}) or (p_{1avg} - p_{2avg}), falls outside of the Region of Certainty and, therefore, into the Region of Uncertainty, the Null Hypothesis is rejected and the Alternate Hypothesis is accepted.

The Critical Value test is much easier to implement with Excel than the p Value test. You will believe it after you review the p Value test below.

b) p Value Test

The p Value Test is an equivalent alternative to the Critical Value Test and also tells whether to reject or not reject the Null Hypothesis.

The p Value equals the percentage of area under the Normal curve that is in the tail outside of the actual value of the variable (p_{avg}) or (p_{1avg} - p_{2avg}).

For a one-tailed test, if the p Value is larger than α, the Null Hypothesis is not rejected. For a two-tailed test, if the p Value is larger than **α**/2, the Null Hypothesis is not rejected.

For a one-tailed test, the Region of Uncertainty is contained entirely in one tail. Therefore the curve area contained by the Region of Uncertainty in that tail equals α.

For a two-tailed test, the Region of Uncertainty is split between both tails. Therefore the curve area contained by the Region of Uncertainty in each tail equals α/2.

There are two possibilities for calculating the p Value. Each possibility depends on whether the actual value of the distributed variable ($\mathbf{p_{avg}}$) or ($\mathbf{p_{1avg}} - \mathbf{p_{2avg}}$) is greater or less than the Normal curve mean, which is established by the Null Hypothesis.

Actual value of the variable is greater (to the right of) the mean.

p $\text{Value}_{\mathbf{pavg}}$ = **1** - NORMSDIST(**[** $\mathbf{p_{avg}} - \mu$ **]** / $\mathbf{s_{pavg}}$)

(**[** $\mathbf{p_{avg}} - \mu$ **]** / $\mathbf{s_{pavg}}$) = number of standard errors the $\mathbf{p_{avg}}$ is from the mean

Excel note - NORMSDIST(x) calculates the total area under the Normal curve to the left of the point that is x standard errors to the right of the Normal curve mean.

Actual value of the variable is less than (to the left of) the mean.

p $\text{Value}_{\mathbf{pavg}}$ = NORMSDIST(**[** $\mathbf{p_{avg}} - \mu$ **]** / $\mathbf{s_{pavg}}$)

(**[** $\mathbf{p_{avg}} - \mu$ **]** / $\mathbf{s_{pavg}}$)= number of standard errors that p_{avg} is from the mean

Excel note - NORMSDIST(x) calculates the total area under the Normal curve to the left of the point that is x standard errors to the right of the Normal curve mean. If x is to the right of the mean, x will be positive and NORMSDIST(x) will be greater than 0.50. If x is to the left of the mean, x will be negative and NORMSDIST(x) will be less than 0.50.

For a two-tailed test---> When the p Value is greater than α/2, the actual value of the distributed variable falls inside the Region of Certainty and the Null Hypothesis is not rejected.

For a one-tailed test---> When the p Value is greater than α, the actual value of the distributed variable falls inside the Region of Certainty and the Null Hypothesis is not rejected.

The p Value test is better understood by examining the example problems that follow. As mentioned before, the Critical Value test is equivalent to the p Value test but is much easier to implement in Excel.

Type 1 and Type 2 Errors

Type 1 Error occurs if the null hypothesis is incorrectly rejected when it is actually true. In other words, it is incorrectly believed that a parameter changed when it did not.

Type 2 Error occurs if the null hypothesis is incorrectly retained when it is actually false. In other words, it is incorrectly believed that a parameter did not change but it did.

Problems

Problem 1 - Two-Tailed, Two Sample, Unpaired Hypothesis Test of Proportion -Testing employee preferences in two companies

Problem: A survey was taken to determine whether two different companies would have the same percentage of workers who would prefer a pay increase over a benefits increase. In a random sample of 150 workers at Company 1, 75 indicated that they would prefer a pay increase. In a random sample of 200 workers at Company 2, 103 of them preferred the pay increase. In each company, the sample was less than 5% of the total number of workers. Determine with the maximum possibility of error of 1% whether the proportions of employees desiring the pay increase are the same at both companies.

We know that this is a **test of proportion** and not mean because **each individual sample taken has only 2 possible values**: An employee prefers either a pay increase or benefits.

We know that this is a **two-tailed test** because we are trying to **determine if the proportions at each plant are merely different**, not whether one proportion is larger or smaller than the other.

We know that **two samples** must be taken because no data is initially available.

This is **unpaired data** because groups are sampled independently. (Proportion testing cannot be applied to Paired data)

Company 1

p_{1avg} = Sample proportion 1= 75/150 = 0.50

q_{1avg} = 1 - p_{1avg} = 1 - 0.50 = 0.50

n1 = Sample size 1 = 150

Company 2

p_{2avg} = Sample proportion 2 = 103/200 = 0.515

q_{2avg} = 1 - p_{2avg} = 1 - 0.515 = 0.485

n2 = Sample size 2 = 200

α = Level of significance = 0.01

---> 1% max chance of error

---> 99% Level of Certainty Required

This problem can be solved using the standard four-step method for Hypothesis testing.

Step 1 - Create the Null and Alternate Hypotheses

The **Null Hypothesis** normally states that both populations sampled are the same. If the proportions p_{1avg} and p_{2avg}, the proportions of each company population that prefer pay raises instead of increased Benefits are the same, then $p_{1avg} - p_{2avg} = 0$.

The Null Hypothesis states that both companies have the same proportions, which is equivalent to:

Null Hypothesis, H_0 ----> $p_{1avg} - p_{2avg} = 0$

The **Alternate Hypothesis** states that the proportions of employees in each company that prefer pay raises instead of increased benefits are different, which is equivalent to: The Alternate Hypothesis, which states that p_{1avg} is different than p_{2avg}, is as follows:

Alternate Hypothesis, H_1 ----> $p_{1avg} - p_{2avg} \neq 0$

For this two-tailed test, the Alternative Hypothesis states that the value of the distributed variable ($p_{1avg} - p_{2avg}$) does not equal the value stated by the Null Hypothesis, which is 0.

The **Region of Uncertainty** will be split between **both outer tails**.

Note - the Alternative Hypothesis determines whether the Hypothesis test is a one-tailed test or a two-tailed test as follows:

One-tailed test ----> (Value of variable) **is greater than** OR **is less than** (Constant)

Two-tailed test ----> (Value of variable) **does not equal** (Constant)

Step 2 - Map the Normal Curve

We now create a Normal curve showing a distribution of the same variable that is used by the Null Hypothesis, which is (p_{1avg} - p_{2avg})
The mean of this Normal curve will occur at the same value of the distributed variable as stated in the Null Hypothesis.

Since the Null Hypothesis states that $p_{1avg} - p_{2avg} = 0$, the Normal curve will map the distribution of the variable (p_{1avg} - p_{2avg}) with a mean of (p_{1avg} - p_{2avg}) = 0

This Normal curve will have a standard error that is calculated as the standard error of a sampled proportion is normally calculated, as follows:

The standard error of the difference of proportions is: $s_{(p1avg - p2avg)}$ =
= SQRT [$p_{weighted}$ * $q_{weighted}$ (1 / n1 + 1 / n2)]

$p_{weighted}$ = (n1* p_{1avg} + n2* p_{2avg}) / (n1 + n2)
= [(150 * 0.50) + (200 * 0.515) / (150 + 200)]
= 0.51

$q_{weighted}$ = 1 - $p_{weighted}$
= 1 - 0.51 = 0.49

Standard Error = $s_{(p1avg - p2avg)}$
= SQRT [(0.51 * 0.49) * (1 / 150 + 1 / 200)]
= 0.054

The Normal curve for this problem is shown on the next page.

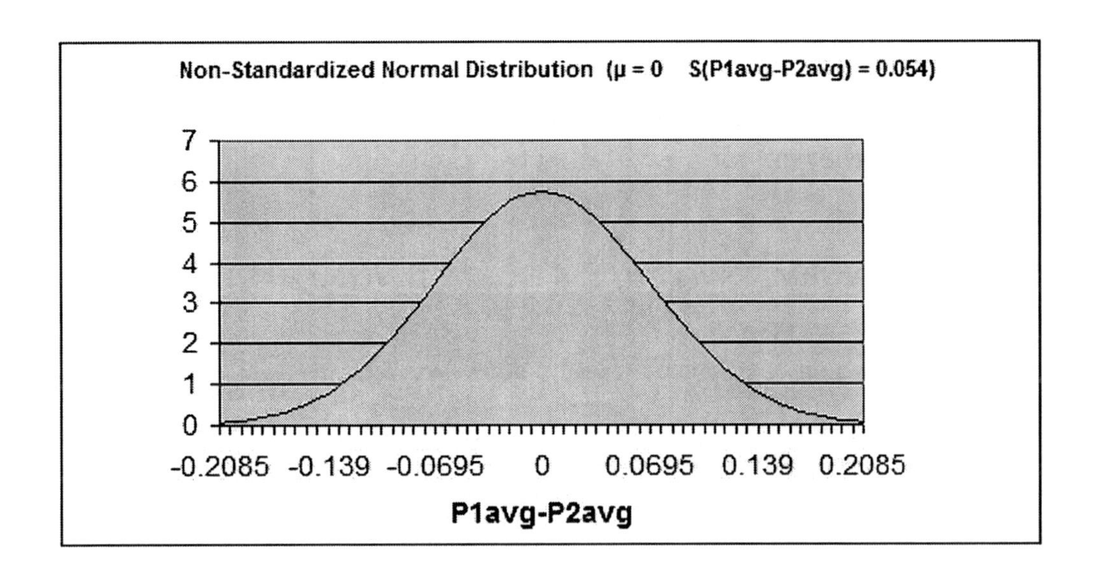

Step 3 - Map the Region of Certainty

The problem requires a 99% Level of Certainty so the Region of Certainty will contain 99% of the area under the Normal curve.

We know that this problem uses a two-tailed test with the Region of Uncertainty split between both outer tails. Each outer tail will contain $\alpha/2$ of the total area, or 0.005 (0.5% of the total area).

The Region of Uncertainty contains 1% of the total area under the Normal curve. The entire **99%** Region of Certainty lies in the center of the Normal curve with 1/2 of the total Region of Uncertainty contained in each outer tail. For a two-tailed test, one tail contains only 1/2 of α. Since $\boldsymbol{\alpha}$ = 0.01 and corresponds to the Level max possibility of error of 1%, each tail contains 0.005 (0.5%) of the total area under the curve.

We need to find out how far the boundary of the Region of Certainty is from the Normal curve mean. Calculating the number of standard errors from the Normal curve mean to the outer boundary of the Region of Certainty in either tail for a two-tailed test tail is done as follows:

$Z_{99\%,2\text{-tailed}}$ = NORMSINV(1 - α/2) = NORMSINV(**0.995**) = 2.58

Excel Note - NORMSINV(x) = The number of standard errors from the Normal curve mean to a point right of the Normal curve mean at which x percent of the area under the Normal curve will be to the left of that point. For a two-tailed test, x equals the Level of Certainty plus 1/2 the Level of Uncertainty. This is because the area under the curve to the left of x will contain the entire Region of Certainty and the 1/2 part of the Region of Uncertainty that exists in the outer left tail.

Additional note - For a two-tailed test, NORMSINV(x) can be used to calculate the number of standard errors from the Normal curve mean to the boundary of the Region of Certainty for each boundary in the left or the right tail. For the two-tailed test, both outer boundaries will be the same distance from the Normal curve mean.

The Region of Certainty extends to the left and to the right of the Normal curve mean of $(p_{1avg} - p_{2avg}) = 0$ by 2.58 standard errors.

One standard error = $s_{(p1avg-p2avg)} = 0.054$, so:

2.58 standard errors = (2.58) * (0.054) = 0.139

The outer boundaries of the Region of Certainty have the values
$= \mu +/- Z_{99\%,2\text{-tailed}} * s_{(p1avg-p2avg)}$

which equals 0 +/- (2.58) * (0.054) = +/- 0.139

These points (-0.139 and 0.139) are 2.58 standard errors from the Normal curve mean of $(p_{1avg} - p_{2avg}) = 0$

These points (-0.139 and 0.139) are the left and right boundaries of the 99% Region of Certainty on the Normal curve.

This is shown in the following diagram:

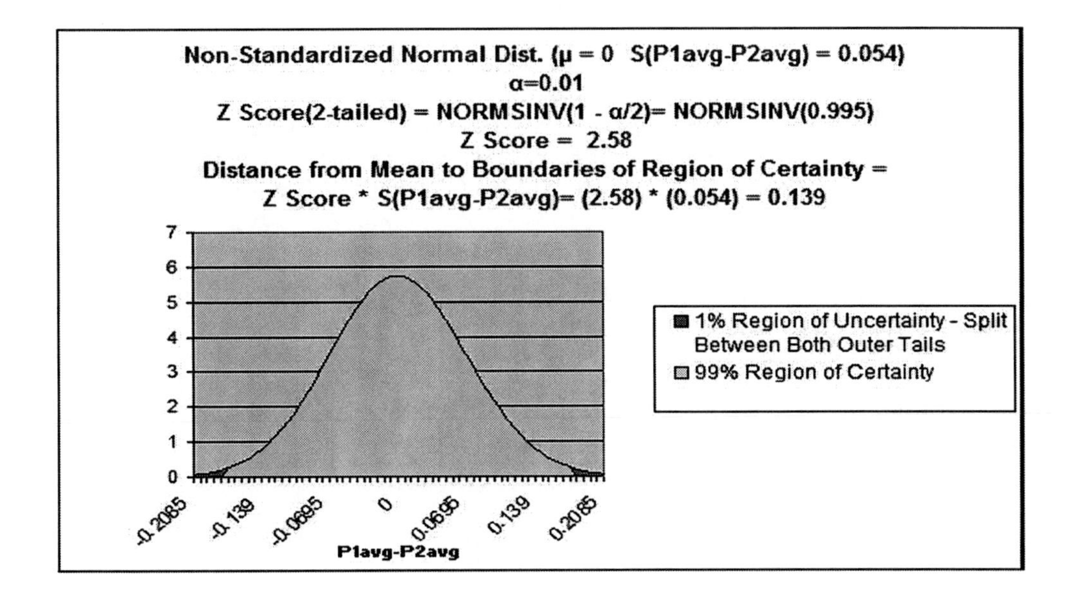

Step 4 - Perform Critical Value and p-Value Tests

a) Critical Value Test

The Critical Value Test is the final test to determine whether to reject or not reject the Null Hypothesis. The p Value Test, described below, is an equivalent alternative to the Critical Value Test.

The Critical Value test tells whether the value of the actual variable, p_{1avg} - p_{2avg}, falls inside or outside of the **Critical Value**, which is the **boundary between the Region of Certainty and the Region of Uncertainty**.

If the actual value of the distributed variable, $\mathbf{p_{1avg} - p_{2avg}}$, falls within the Region of Certainty, the Null Hypothesis is not rejected.

If the actual value of the distributed variable, $\mathbf{p_{1avg} - p_{2avg}}$, falls outside of the Region of Certainty and, therefore, into the Region of Uncertainty, the Null Hypothesis is rejected and the Alternate Hypothesis is accepted.

In this case, the actual value of the variable ($\mathbf{p_{1avg} - p_{2avg}}$)= 0.50 - 0.515 = -0.015

The actual value of the variable ($\mathbf{p_{1avg} - p_{2avg}}$) = -0.015 and is therefore inside of (to the right of) the outer left Critical Value (-0.139), which is the boundary between the Regions of Certainty and Uncertainty in the left tail.

The actual value of the variable ($\mathbf{p_{1avg} - p_{2avg}}$) is inside the Region of Certainty and therefore inside the Critical Value.

We therefore do not reject the Null Hypothesis and state that it is not disproved that the proportions of employees in each company who prefer pay increase over benefits are statistically the same.

This is shown in the following diagram on the next page.

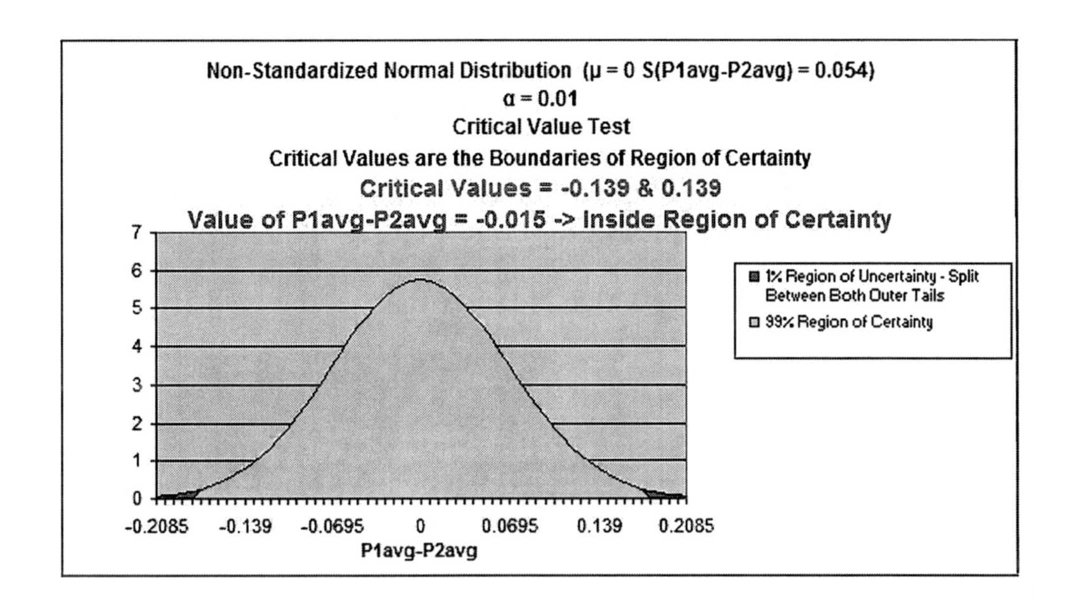
Non-Standardized Normal Distribution (μ = 0 S(P1avg-P2avg) = 0.054)
α = 0.01
Critical Value Test
Critical Values are the Boundaries of Region of Certainty
Critical Values = -0.139 & 0.139
Value of P1avg-P2avg = -0.015 -> Inside Region of Certainty
1% Region of Uncertainty - Split Between Both Outer Tails
99% Region of Certainty
P1avg-P2avg
-0.2085
-0.139
-0.0695
0
0.0695
0.139
0.2085

b) p Value Test

The p Value Test is an equivalent alternative to the Critical Value Test and also tells whether to reject or not reject the Null Hypothesis. The p Value equals the percentage of area under the Normal curve that is in the tail outside of the actual value of the variable **(p_{1avg} - p_{2avg})**.

For a one-tailed test, if the p Value is larger than **α**, the Null Hypothesis is not rejected. For a two-tailed test, if the p Value is larger than **α**/2, the Null Hypothesis is not rejected.

For a one-tailed test, the Region of Uncertainty is contained entirely in one tail. Therefore the curve area contained by the Region of Uncertainty in that tail equals α.

For a two-tailed test, the Region of Uncertainty is split between both tails. Therefore the curve area contained by the Region of Uncertainty in that tail equals α/2.

The p Value for the actual value of the distributed variable, which in this case is less than the mean (falls to the left of the mean), is: p $\text{Value}_{(p1avg-p2avg)}$ = NORMSDIST(**[($p1_{avg}$ - $p2_{avg}$) - μ]** / **s(p1avg-p2avg)**)

Excel note - NORMSDIST(x) calculates the total area under the Normal curve to the left of the point that is x standard errors to the right of the Normal curve mean. If x is to the right of the mean, x will be positive and NORMSDIST(x) will be greater than 0.50. If x is to the left of the mean, x will be negative and NORMSDIST(x) will be less than 0.50.

p $\text{Value}_{(p1avg-p2avg)}$

= NORMSDIST((-0.015 - 0) / 0.054)
= NORMSDIST(-0.015/0.054)
= 0.39

The p Value (0.39) is greater than α/2 (0.005), so the Null Hypothesis is not rejected.

For a two-tailed test---> When the p Value is greater than α/2, the actual of the distributed variable falls inside the Region of Certainty and the Null Hypothesis is not rejected.

This is the case here for this two-tailed test and is shown as follows.

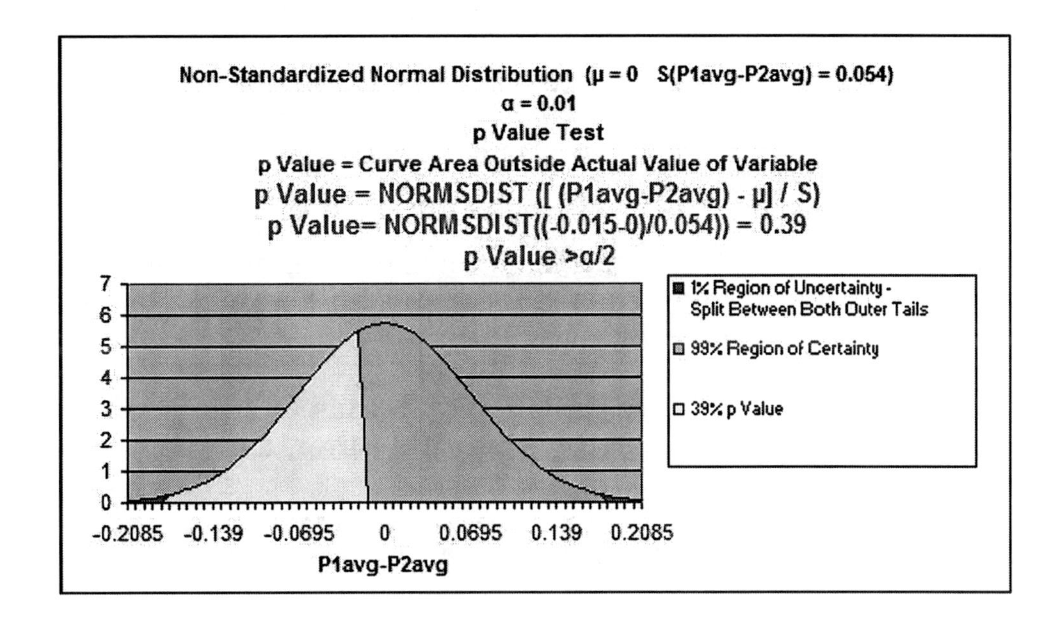

Problem 2 - One-Tailed, Two Sample, Unpaired Hypothesis Test of Proportion - Testing effectiveness of two drugs

Problem: A pharmaceutical manufacturer is testing the effectiveness of two different drugs to cure the same illness. The pharmaceutical company is trying to determine whether Drug 2 is better than Drug 1. The first drug was administered to 80 patients and cured 52 of them. The 2nd drug was administered to 90 patients and cured 63 of them. Determine within an error of 1% whether Drug 2 was more effective than Drug 1.

We know that this is a **test of proportion** and not mean because **each individual sample taken has only 2 possible values**: the drug either cured the patient or it didn't.

We know that this is a **one-tailed test** because we are trying to **determine if one drug is better than the other**, not whether they merely have different results.

We know that **two samples** must be taken because no data is initially available.

This is **unpaired data** because groups are sampled independently. (Proportion testing cannot be applied to Paired data)

Drug 1

$\mathbf{p_{1avg}}$ = Sample proportion 1 = 52 / 80 = 0.65 cured

$\mathbf{q_{1avg}}$ = 1 - $\mathbf{p_{1avg}}$ = 1 - 0.65 = 0.35 not cured

n1 = Sample size 1 = 80

Drug 2

$\mathbf{p_{2avg}}$ = Sample proportion 2 = 63 / 90 = 0.70 cured

$\mathbf{q_{1avg}}$ = 1 - $\mathbf{p_{1avg}}$ = 1 - 0.70 = 0.30 not cured

n2 = Sample size 2 = 90

α = 0.01 = alpha = Level of significance
---> 1% max chance of error
---> 99% Level of Certainty Required

This problem can be solved using the standard four-step method for Hypothesis testing.

Step 1 - Create the Null and Alternate Hypotheses

The **Null Hypothesis** normally states that both populations sampled are the same. If the proportions $\mathbf{p_{1avg}}$ and **p**2avg, cured from each drug are the same, then $\mathbf{p_{1avg}}$ - $\mathbf{p_{2avg}}$ = 0.

The Null Hypothesis states that both drugs have the same effectiveness, which is equivalent to:

Null Hypothesis = H_o = $\mathbf{p_{1avg}}$ - $\mathbf{p_{2avg}}$ = 0

The **Alternate Hypothesis** states that Drug 2 is better than Drug 1, which is equivalent to:

The Alternate Hypothesis, which states that p_{2avg} is greater than p_{1avg}, is as follows:

Alternate Hypothesis, $H_1 = p_{1avg} - p_{2avg} < 0$

For this one-tailed test, the Alternative Hypothesis states that if the value of the distributed variable ($p_{1avg} - p_{2avg}$) is less than the value stated by the Null Hypothesis, the **Region of Uncertainty** will be in the outer **left tail**.

Note - the Alternative Hypothesis determines whether the Hypothesis test is a one-tailed test or a two-tailed test as follows:

One-tailed test ----> (Value of variable) **is greater than** **OR is less than** (Constant)

Two-tailed test ----> (Value of variable) **does not equal** (Constant)

Step 2 - Map the Normal Curve

We now create a Normal curve showing a distribution of the same variable that is used by the Null Hypothesis, which is ($p_{1avg} - p_{2avg}$)

The mean of this Normal curve will occur at the same value of the distributed variable as stated in the Null Hypothesis.

Since the Null Hypothesis states that $p_{1avg} - p_{2avg} = 0$, the Normal curve will map the distribution of the variable ($p_{1avg} - p_{2avg}$) with a mean of ($p_{1avg} - p_{2avg}$) = 0

This Normal curve will have a standard error that is calculated as the standard error of a sampled proportion is normally calculated, as follows:

The standard error of the difference of proportions is:
$s_{(p1avg-p2avg)}$ = SQRT [$p_{weighted}$ * $q_{weighted}$ (1 / n1 + 1 / n2)]

$p_{weighted}$ = (n1*p_{1avg} + n2*p_{2avg}) / (n1 + n2)
= [(80 * 0.65) + (90 * 0.70) / (80 + 90)]
= 0.676

$q_{weighted}$ = 1 - $p_{weighted}$
= 1 - 0.676
= 0.324

Standard Error = $s_{(p1avg-p2avg)}$

= SQRT [$p_{weighted}$ * $q_{weighted}$ (1 / n1 + 1 / n2)]

= SQRT [(0.676 * 0.324) * (1 / 80 + 1 / 90)]

= 0.0719

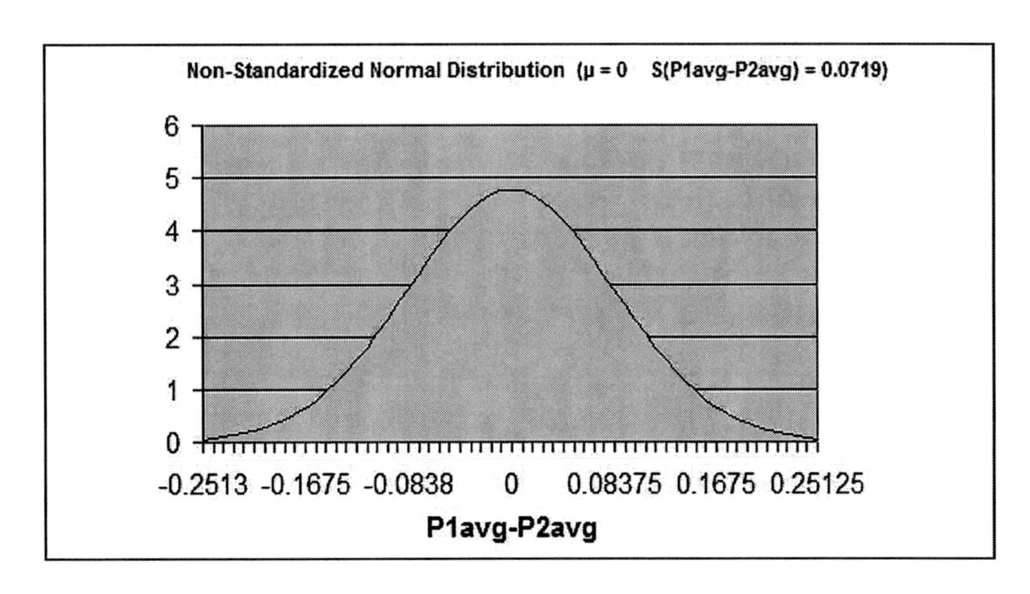

Step 3 - Map the Region of Certainty

The problem requires a 99% Level of Certainty so the Region of Certainty will contain 99% of the area under the Normal curve.

We know that this problem uses a one-tailed test with the Region of Uncertainty contained entirely in the outer left tail.

The Region of Uncertainty contains 1% of the total area under the Normal curve. The entire **99%** Region of Certainty lies to the right of the one and only Region of Uncertainty in the outer left tail. For a one-tailed test, one tail contains the entire α, or total Region of Uncertainty. In this case, $\alpha = 0.01$, which is **1%** of the total curve area.

We need to find out how far the boundary of the Region of Certainty is from the Normal curve mean. Calculating the number of standard errors from the Normal curve mean to the outer boundary of the Region of Certainty in the left tail is done as follows:

$Z_{99\%,1\text{-tailed}}$ = NORMSINV(1 - α) = NORMSINV(**0.99**) = 2.33

Excel Note - NORMSINV(x) = The number of standard errors from the Normal curve mean to a point right of the Normal curve mean at which x percent of the area under the Normal curve will be to the left of that point. x equals the Level of Certainty required by the one-tailed problem.

Additional note - For a one-tailed test, NORMSINV(x) can be used to calculate the number of standard errors from the Normal curve mean to the boundary of the Region of Certainty whether the boundary is in the left or the right tail.

The Region of Certainty extends to the left of the Normal curve mean of (p_{1avg} - p_{2avg}) = 0 by 2.33 standard errors.

One standard error = $s_{(p1avg-p2avg)}$ = 0.0719, so: 2.33 standard errors = (2.33) * (0.0719) = 0.1675

The outer left boundary of the Region of Certainty has the value

= μ - $Z_{99\%,1\text{-tailed}}$ * $s_{(p1avg-p2avg)}$
which equals 0 - (2.33) * (0.0719) = - 0.1675

This point (-0.1675) is 2.33 standard errors to the left of the Normal curve mean of (p_{1avg} - p_{2avg}) = 0

This point (-0.1675) is the left boundary of the 99% Region of Certainty on the Normal curve.

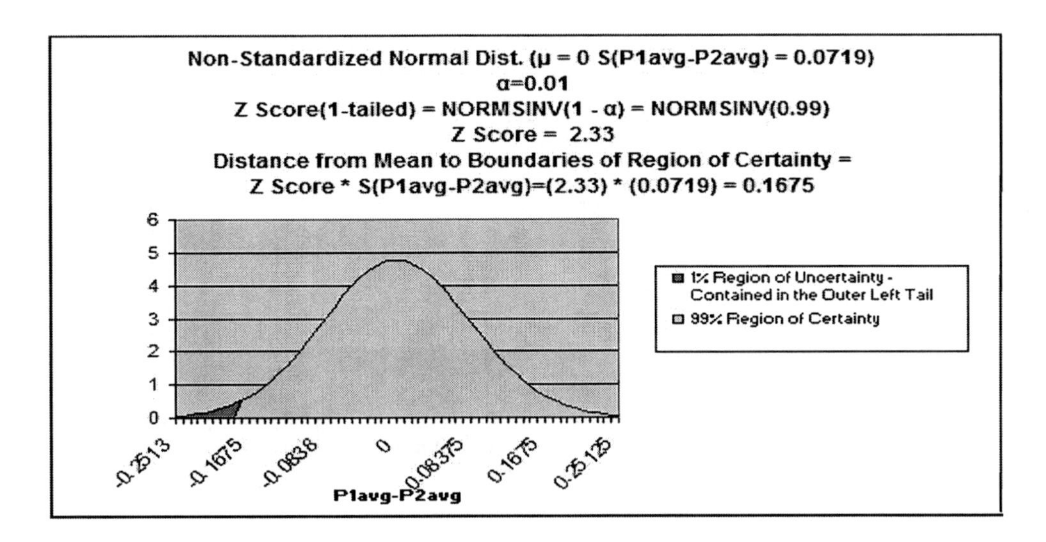

Step 4 - Perform Critical Value and p-Value Tests

a) Critical Value Test

The Critical Value Test is the final test to determine whether to reject or not reject the Null Hypothesis. The p Value Test, described below, is an equivalent alternative to the Critical Value Test.

The Critical Value test tells whether the value of the actual variable, $p_{1avg} - p_{2avg}$, falls inside or outside of the **Critical Value**, which is the **boundary between the Region of Certainty and the Region of Uncertainty**.

If the actual value of the distributed variable, $p_{1avg} - p_{2avg}$, falls within the Region of Certainty, the Null Hypothesis is not rejected.

If the actual value of the distributed variable, $p_{1avg} - p_{2avg}$, falls outside of the Region of Certainty and, therefore, into the Region of Uncertainty, the Null Hypothesis is rejected and the Alternate Hypothesis is accepted.

In this case, the actual value of the variable, $p_{1avg} - p_{2avg}$ = = 0.65 - 070 = -0.05

The actual value of the variable ($p_{1avg} - p_{2avg}$) = -0.05 and is therefore inside the Critical Value (-0.1675), which is the boundary between the Regions of Certainty and Uncertainty.

The actual value of the variable ($p_{1avg} - p_{2avg}$) is inside the Region of Certainty and therefore inside the Critical Value.

We therefore do not reject the Null Hypothesis and state that It is not disproven that the proportions of patients cured by each drug are statistically the same.

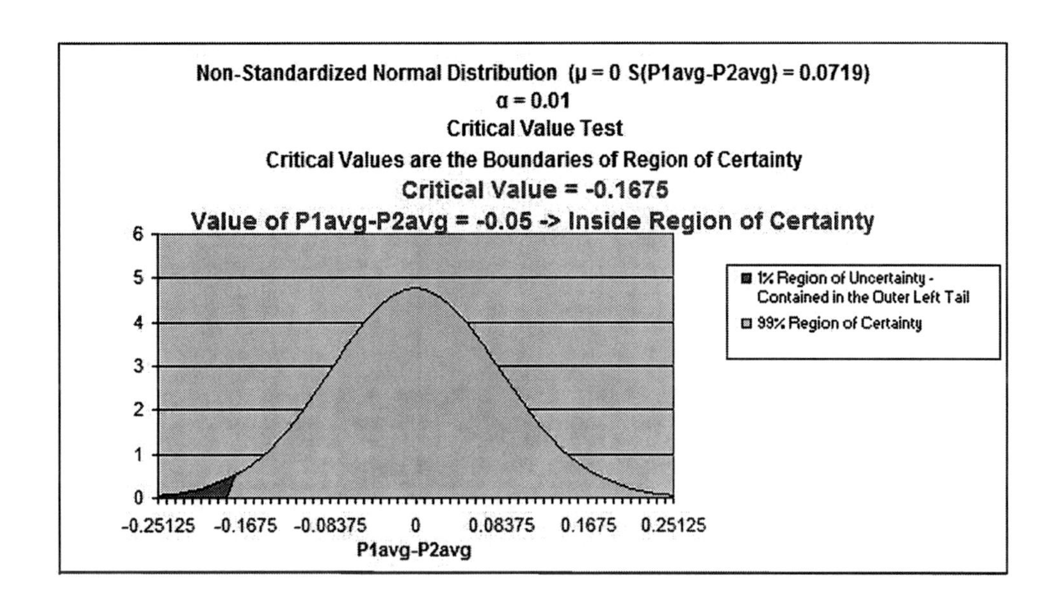

b) p Value Test

The p Value Test is an equivalent alternative to the Critical Value Test and also tells whether to reject or not reject the Null Hypothesis.

The p Value equals the percentage of area under the Normal curve that is outside of the actual value of the variable ($\mathbf{p_{1avg}}$ - $\mathbf{p_{2avg}}$).

For a one-tailed test, if the p Value is larger than α, the Null Hypothesis is not rejected. For a two-tailed test, if the p Value is larger than **α/2**, the Null Hypothesis is not rejected.

For a one-tailed test, the Region of Uncertainty is contained entirely in one tail. Therefore the curve area contained by the Region of Uncertainty in that tail equals α.

For a two-tailed test, the Region of Uncertainty is split between both tails. Therefore the curve area contained by the Region of Uncertainty in that tail equals α/2.

The p Value for the actual value of the distributed variable, which in this case is less than the mean (falls to the left of the mean in the left tail), is:

p Value$_{(p1avg-p2avg)}$ = NORMSDIST(**[** (**p$_{1avg}$** - **p$_{2avg}$**) - **µ** **]** / **s$_{(p1avg-p2avg)}$**)

Excel note - NORMSDIST(x) calculates the total area under the Normal curve to the left of the point that is x standard errors to the right of the Normal curve mean.

Please note that if the p Value in the case that the actual value of the distributed variable were to the right of the mean, the p Value would be calculated as follows:

p Value$_{(p1avg-p2avg)}$ = **1** - NORMSDIST(**[** (**p$_{1avg}$** - **p$_{2avg}$**) - **µ** **]** / **s$_{(p1avg-p2avg)}$**)

p Value$_{(p1avg-p2avg)}$ = NORMSDIST((-0.05 - 0) / 0.0719)

= NORMSDIST(-0.05/0.0719) = 0.24

The p Value (0.24) is greater than α (0.01), so the Null Hypothesis is not rejected.

For a one-tailed test---> When the p Value is greater than α, the actual value of the distributed variable falls inside the Region of Certainty and the Null Hypothesis is not rejected.

This is the case here as shown in the following diagram on the next page.

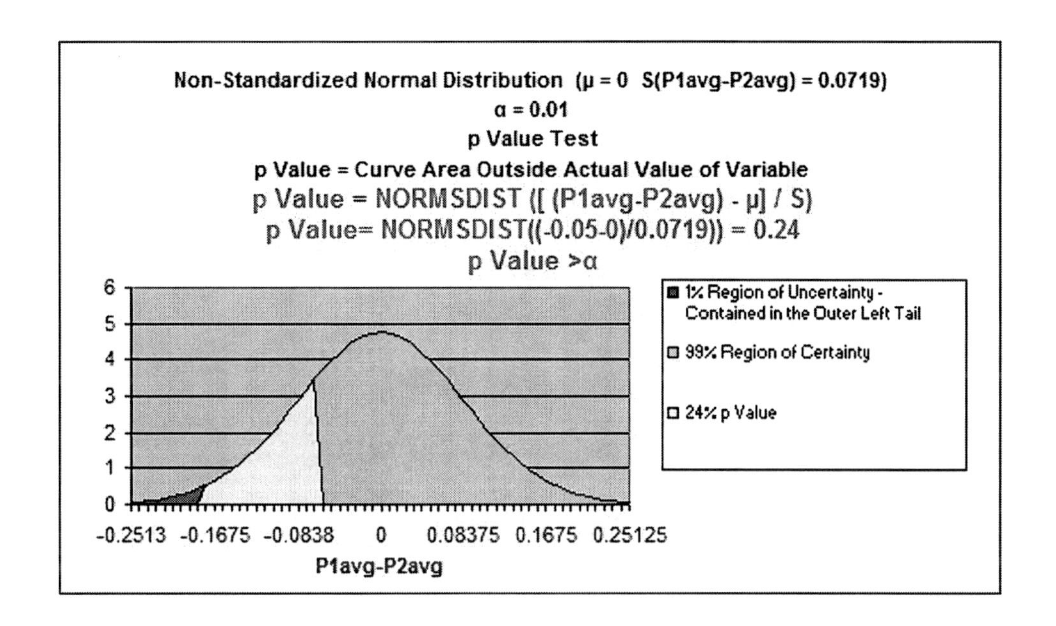
Non-Standardized Normal Distribution (μ = 0 S(P1avg-P2avg) = 0.0719)
α = 0.01
p Value Test
p Value = Curve Area Outside Actual Value of Variable
p Value = NORMSDIST ([(P1avg-P2avg) - μ] / S)
p Value= NORMSDIST((-0.05-0)/0.0719)) = 0.24
p Value >α
6
5
4
3
2
1
0
-0.2513 -0.1675 -0.0838 0 0.08375 0.1675 0.25125
P1avg-P2avg
1% Region of Uncertainty - Contained in the Outer Left Tail
99% Region of Certainty
24% p Value

Chapter 10 - Excel Hypothesis Tools

Excel offers several data analysis tools that are designed to be used for Hypothesis testing and are described in this module. They are:

t-Test: Paired Two Sample for Means

t-Test: Two-Sample Assuming Equal Variances

t-Test: Two-Sample Assuming Unequal Variances

z-Test: Two Sample for Means

ZTEST

TTEST

The first four tools can be found at:

Tools / Data Analysis (in Excel 2010 -> Data tab)

ZTEST and **TTEST** are statistical functions found at:

Insert / Function / Statistical

One Caution About Using t-Tests on Small Samples

We will explain exactly how each of these tools is used below. One caution with using any of the t-Test tools:

t-Tests are often performed when only small samples (n<30) are available. This should only be done if it can be shown that the underlying population from which the samples are drawn is reasonably Normally distributed. The distribution of the underlying data population should be evaluated first before performing any statistical analysis.

One quick and easy way to analyze your data in Excel for Normality is to create a histogram of that data. If you see that the histogram shape reasonably resembles a Normal curve, you are probably safe to apply statistical tools which require that the data population to be normally distributed, such as t-distribution tests. If any t-Test is applied to a small sample drawn from a population that is not normally distributed, the resulting analysis can be totally wrong.

Nonparametric tests should be applied when the data is not normally distributed. There is a work-around that can be applied when the data population is not normally distributed. See the course module about Confidence Intervals for further details. As a final result, nonparametric tests can be applied when the underlying population is not normally distributed. Nonparametric tests are discussed in a different manual of the Excel Statistical Master series.

Statistical analysis can be applied to large samples using t-Tests. The t distribution approaches the Normal distribution as sample size becomes larger. Statistical tests using the Normal distribution can be applied to large samples (n>30) due to the Central Limit Theorem. Further explanation of this theorem can be found in the course module entitled Confidence Intervals.

Having said all that, here are the three t-Tests and one z-Test that are convenient, built-in statistical analysis tools in Excel:

t-Test: Paired Two Sample for Means

Problem: Google and Yahoo each provide two types of networks that advertisers can run pay-per-click ads on. One type of network is the search network and the other is a content network. One advertiser is running all of his ads exclusively on the search networks of both Google and Yahoo. This advertiser wants to determine with 95% accuracy whether advertising on the content networks of Google and Yahoo will produce more clicks than advertising on the search networks. The advertiser simultaneously ran the same ad on the search and content networks of both Google and Yahoo. This procedure was repeated with 15 different ads. In the table below are the clicks from all tests.

The data is shown on the following page:

Ad	Search Engine	Search Network	Content Network
1	Google	37,663	31,909
	Yahoo	42,345	41,208
2	Google	31,104	38,816
	Yahoo	41,233	43,306
3	Google	32,903	35,375
	Yahoo	42,657	52,354
4	Google	29,829	30,886
	Yahoo	39,612	49,428
5	Google	34,625	38,727
	Yahoo	42,651	43,236
6	Google	31,923	34,565
	Yahoo	39,999	43,865
7	Google	37,664	31,902
	Yahoo	42,342	41,204
8	Google	31,105	38,813
	Yahoo	41,239	43,305
9	Google	32,904	35,372
	Yahoo	42,658	52,353
10	Google	29,823	30,886
	Yahoo	39,616	49,425
11	Google	34,622	38,724
	Yahoo	42,650	43,233
12	Google	31,924	34,565
	Yahoo	39,995	43,862
13	Google	29,826	30,888
	Yahoo	39,617	49,427
14	Google	34,628	38,726
	Yahoo	42,659	43,235
15	Google	31,920	34,564
	Yahoo	39,998	43,863

This is a Hypothesis test because we are determining whether the "After Data" is different in either direction or different in only one direction than the "Before" data. We know that this will require a one-tailed test because we are being asked to determine if the "After Data" is different in just one direction (greater) than the "Before" data.

This is classified by Excel as “paired data” because the test is being duplicated on both search engines.

This is a two-sample test because no initial data was available. Samples had to be taken for each of the two means being tested.

The problem requests a 5% Level of Significance which equates to a 95% Level of Certainty. The Level of Significance is often referred to as α (alpha) so in this case, α = 0.05.

The Null Hypothesis for this problem would state that there is no difference between the number of clicks that the ads generate on the search or content networks.

Expressed another way, the mean difference between the number of clicks on both types of networks is 0.

The Alternate Hypothesis states that ads on the content networks generate more clicks.

This Excel tool is located at:

Tools / Data Analysis / t-Test: Paired Two Sample for Means (in Excel 2010 -> Data tab)

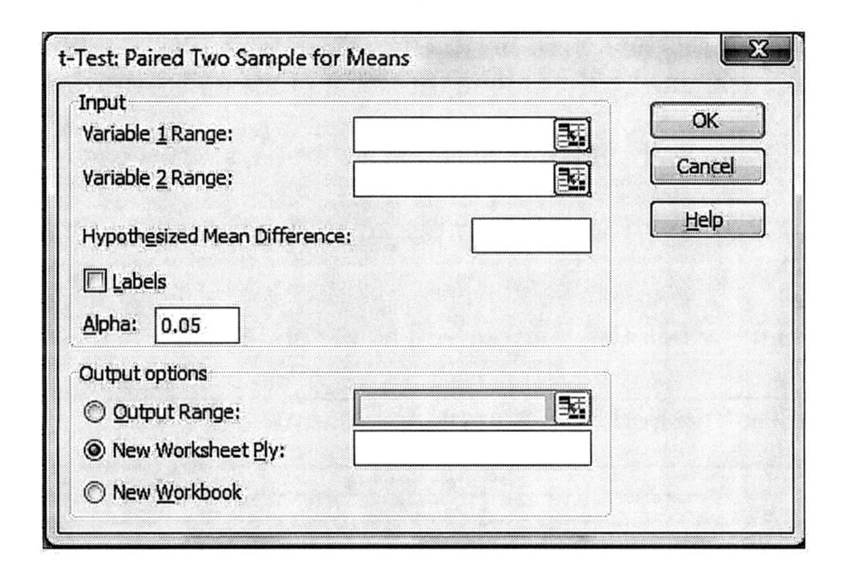

The tool's dialog box asks for the following 6 pieces of data:

1) **Variable 1 Range**: Variable 1 represents the "Before Data." In this case, the "Before Data" is the Search Network data column highlighted in yellow. "Before Data" would have to be arranged in a column and highlighted in this manner, including the column label on the top row.

2) **Variable 2 Range**: Variable 2 represents the "After Data." In this case, the "After Data" is the Content Network data column highlighted in tan. "After Data" would have to be arranged in a column and highlighted in this manner, including the column label on the top row.

3) **Hypothesized Mean Difference**: This is normally 0 for a Hypothesis test of paired data. The Hypothesized Mean Difference will correspond to whatever is on the right side of the equal sign for the Null Hypothesis.

4) **Includes Labels?** Yes, if you included them when the columns were highlighted, No, if they were not included when the data columns were highlighted. You must either highlight and include all labels or no labels.

5) **Alpha** - The Level of Significance equals alpha, so α = 0.05

6) **Output Range** - The output range must have at least 14 rows and 3 columns free.

Click OK and the following output will be generated:

t-Test: Paired Two Sample for Means

	Search Network	*Content Network*
Mean	**37,058**	**40,267**
Variance	22,513,133	41,031,253
Observations	30	30
Pearson Correlation	0.76	
Hypothesized Mean Difference	0	
df	29	
t Stat	-4.25	
P(T<=t) one-tail	**0.00010**	
t Critical one-tail	1.70	
P(T<=t) two-tail	0.00	
t Critical two-tail	2.05	

Interpretation of the data is very simple: Compare the p-Value with the Level of Significance (α). If the p-Value is less than α, then the Null Hypothesis is rejected and the Alternative Hypothesis is accepted.

This is a one-tailed test so the p-value shown for the one-tailed testis the correct p-Value. In this case, the one-tailed p-Value is 0.00010, which is less than the level of significance (α = 0.05) so the Null Hypothesis is rejected, and the Alternate Hypothesis is accepted, which states that the content network produces more clicks than the search network.

t-Test: Two Sample Assuming Unequal Variance

Problem: Evaluate the returns of these two stocks to determine if there is a real difference. Use a 0.05 level of significance.

IBM	Apple
0.7541	-4.6296
14.9701	18.986
11.9792	-1.7226
7.907	-0.5535
-5.1724	6.679
3.4091	1.8261
0.7541	-4.6296
14.9701	18.986
11.9792	-1.7226
7.907	-0.5535
-5.1724	6.679
3.4091	1.8261
0.7541	-4.6296
14.9701	18.986
11.9792	-1.7226
7.907	-0.5535
-5.1724	6.679
3.4091	1.8261
3.4091	1.8261
0.7541	-4.6296
14.9701	18.986
11.9792	-1.7226
7.907	-0.5535
-5.1724	6.679
3.4091	1.8261
0.7541	-4.6296
14.9701	18.986
11.9792	-1.7226
7.907	-0.5535
-5.1724	6.679

This is a Hypothesis test because we are determining whether the mean of one population is different in either direction or different in only one direction than the mean of another population. We know that this will require a two-tailed test because we are being asked to determine if the population means are simply different.

This is a two-sample test because no initial data was available. Samples had to be taken for each of the two means being tested.

We do not initially know the variances of each data set so we Assume them to be unequal. For that reason, this is classified as an Unequal Variance test.

The problem requests a 5% Level of Significance which equates to a 95% Level of Certainty. The Level of Significance is often referred to as α (alpha) so in this case, α = 0.05.

The Null Hypothesis for this problem would state that there is no difference between the average means of both independent populations. Expressed another way, the mean difference between the means = 0.

The Alternate Hypothesis states that the difference between the average means does not equal 0.

On the following page is the Excel dialogue box used to solve the problem:

The tool is located at:

Tools / Data Analysis / t-Test: Two Sample Assuming Unequal Variances (in Excel 2010 -> Data tab)

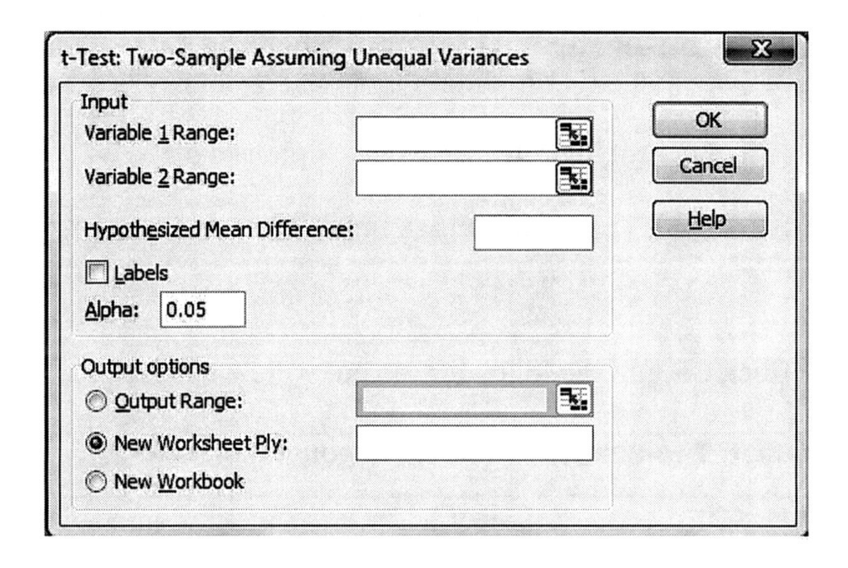

The tool's dialog box asks for the following 6 pieces of data:

1) **Variable 1 Range**: Variable 1 represents Population 1 Data In this case, the Population 1 Data is the IBM return column highlighted in yellow. Population 1 Data would have to be arranged in a column and highlighted in this manner, including the column label on the top row.

2) **Variable 2 Range**: Variable 2 represents Population 2 Data. In this case, the Population 2 Data is the Apple return column highlighted in tan. Population 2 Data would have to be arranged in a column and highlighted in this manner, including the column label on the top row.

3) **Hypothesized Mean Difference**: This is normally 0 for a Hypothesis test of paired data. The Hypothesized Mean Difference will correspond to whatever is on the right side of the equal sign for the Null Hypothesis.

4) **Includes Labels?** Yes, if you included them when the columns were highlighted, No, if they were not included when the data columns were highlighted. You must either highlight and include all labels or no labels.

5) **Alpha** - The Level of Significance equals alpha, so α = 0.05

6) **Output Range** - The output range must have at least 14 rows and 3 columns free.

Click OK and the following output will be generated:

t-Test: Two-Sample Assuming Unequal Variances

	IBM	*Apple*
Mean	**5.64**	**3.43**
Variance	47.95	62.50
Observations	**30**	**30**
Hypothesized Mean Differenc	**0**	
df	57	
t Stat	1.15	
P(T<=t) one-tail	0.13	
t Critical one-tail	1.67	
P(T<=t) two-tail	**0.25**	
t Critical two-tail	2.00	

Interpretation of the data is very simple: Compare the p-Value with the Level of Significance (α). If the p-Value is less than α, then the Null Hypothesis is rejected and the Alternate Hypothesis is accepted.

This is a two-tailed test so the p-value shown for the two-tailed test is the correct p-Value. In this case, the two-tailed p-Value is 0.25, which is greater than the level of significance (α = 0.05) so the Null Hypothesis is not rejected and it hasn't been disproven that the means of the two populations are the same.

t-Test: Two Sample Assuming Equal Variances

This test assumes that the variances of the populations being compared are equal. Usually it is the case that the variances are not known. This test would only be valid for comparing populations of known variances.

Problem: A company is testing batteries from 2 suppliers. Below are listed the hours of usage before each sample ran out. Determine using a 0.05 level of significance whether the new supplier's batteries really last longer than the old supplier's. The variances of battery life of batteries from both suppliers are known to be equal. The following page provides the sample data:

Battery Suppliers

Old Supplier	New Suppliers
61	51
49	42
56	37
51	45
38	65
44	47
61	51
51	75
50	49
60	56
39	52
51	49
43	69
37	53
45	51
65	61
51	49
46	56
50	49
44	58
26	52
52	48
43	79
46	42
64	46
53	52
38	51
44	61
61	49
44	56
42	52
46	49
64	
53	
38	

This is a Hypothesis test because we are determining whether "After Data" is different in either direction or different in only one direction than the "Before" data. We know that this will require a one-tailed test because we are being asked to determine if the "After Data" is different in just one direction (greater) than the "Before" data.

This is a two-sample test because no initial data was available. Samples had to be taken for each of the two means being tested.

The problem requests a 5% Level of Significance which equates to a 95% Level of Certainty. The Level of Significance is often referred to as α (alpha) so in this case, α = 0.05.

The Null Hypothesis for this problem would state that there is no difference between the mean lifetimes of batteries from the old or new suppliers. Expressed another way, the mean difference between mean battery lifetime from the old supplier and from the new supplier is 0.

The Alternate Hypothesis states that batteries from the new supplier last longer. The following page provides the dialogue box use to solve the problem.

The tool is located at:

Tools / Data Analysis / t-Test: Two-Sample Assuming Equal Variances
(in Excel 2010 -> Data tab)

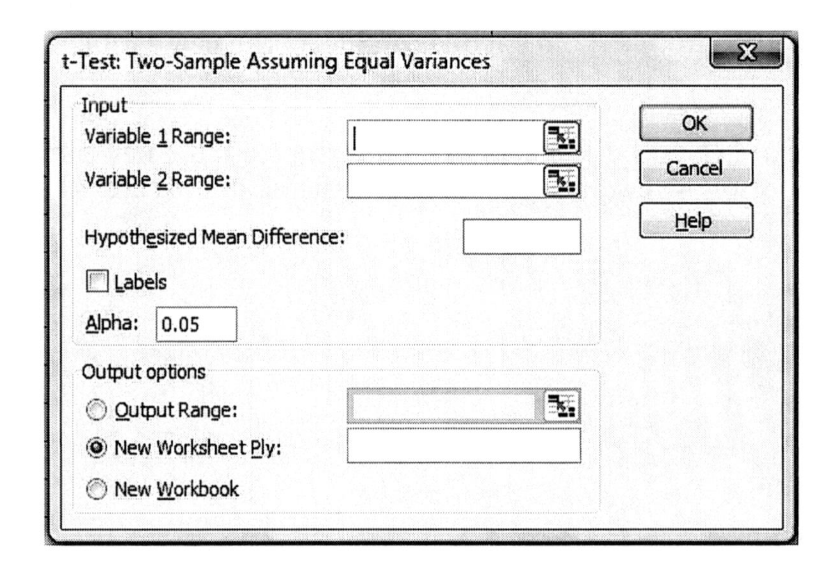

The tool's dialog box asks for the following 6 pieces of data:

1) **Variable 1 Range**: Variable 1 represents the "Before Data." In this case, the "Before Data" is the Old Supplier data column highlighted in yellow. "Before Data" would have to be arranged in a column and highlighted in this manner, including the column label on the top row.

2) **Variable 2 Range**: Variable 2 represents the "After Data." In this case, the "After Data" is the New Supplier data column highlighted in tan. "After Data" would have to be arranged in a column and highlighted in this manner, including the column label on the top row.

3) **Hypothesized Mean Difference**: This is normally 0 for a Hypothesis test of paired data. The Hypothesized Mean Difference will correspond to whatever is on the right side of the equal sign for the Null Hypothesis.

4) **Includes Labels?** Yes, if you included them when the columns were highlighted, No, if they were not included when the data columns were highlighted. You must either highlight and include all labels or no labels.

5) **Alpha** - The Level of Significance equals alpha, so α = 0.05

6) **Output Range** - The output range must have at least 14 rows and 3 columns free.

Click OK and the following output will be generated:

t-Test: Two-Sample Assuming Equal Variances

	Old Suppliers	*New Suppliers*
Mean	**48.74**	**53.19**
Variance	81.31	81.51
Observations	35	32
Pooled Variance	81.41	
Hypothesized Mean [	0	
df	65	
t Stat	-2.01	
P(T<=t) one-tail	**0.02**	
t Critical one-tail	1.67	
P(T<=t) two-tail	0.05	
t Critical two-tail	2.00	

Interpretation of the data is very simple: Compare the p-Value with the Level of Significance (α). If the p-Value is less than α, then the Null Hypothesis is rejected and the Alternate Hypothesis is accepted. This is a one-tailed test so the p-value shown for the one-tailed test is the correct p-Value. In this case, the one-tailed p-Value is 0.02, which is less than the level of significance (α = 0.05) so the Null Hypothesis is rejected and the Alternate Hypothesis is accepted, which states that the batteries from the new supplier have a longer average life.

z-Test: Two Sample for Means

I recommend the use of this z-Test data analysis tool, provided that all sample sizes are large (n>30) and normally distributed. If the underlying distribution is not normally distributed, you can run the z-Test on the means of multi-point samples. For example, if each sample group contains 3 sample points taken randomly and representatively from the underlying population, the means of the sample groups will be normally distributed. Statistics; most basic theorem, the Central Limit Theorem, states that sample means will be Normally distributed no matter how the underlying population is distributed, as long as sample size is large (n>30). You should taken at least 30 sample groups.

You can actually run an experiment in Excel to verify this. Use Excel's random number generator and generate 1,000 random numbers between 0 and 1. Break the set of 1,000 random numbers into random groups of 5 in each group. You will have 200 groups. Take the mean (average of each group). Create a histogram in Excel of those means. The shape of the histogram will be that of a normal curve, no matter how the entire population of 1,000 numbers were distributed.

Here we will apply the Excel data analysis tool **z-Test: Two Sample for Means(in Excel 2010 -> Data tab)** to the same problem as above, since each of the two samples is large (n>30).

Problem: A company is testing batteries from 2 suppliers. Below are listed the hours of usage before each sample stopped working. Determine using a 0.05 level of significance whether the new supplier's batteries really last longer than the old supplier's.

Battery Suppliers

Old Supplier	New Suppliers
61	51
49	42
56	37
51	45
38	65
44	47
61	51
51	75
50	49
60	56
39	52
51	49
43	69
37	53
45	51
65	61
51	49
46	56
50	49
44	58
26	52
52	48
43	79
46	42
64	46
53	52
38	51
44	61
61	49
44	56
42	52
46	49
64	
53	
38	

This is a Hypothesis test because we are determining whether "After Data" is different in either direction or different in only one direction than the "Before" data. We know that this will require a one-tailed test because we are being asked to determine if the "After Data" is different in just one direction (greater) than the "Before" data.

This is a two-sample test because no initial data was available. Samples had to be taken for each of the two means being tested.

The problem requests a 5% Level of Significance which equates to a 95% Level of Certainty. The Level of Significance is often referred to as α (alpha) so in this case, α = 0.05.

The Null Hypothesis for this problem would state that there is no difference between the mean lifetimes of batteries from the old or new suppliers. Expressed another way, the mean difference between mean battery lifetime from the old supplier and from the new supplier is 0.

The Alternate Hypothesis states that batteries from the new supplier last longer.

This is unpaired data with a large number of samples (>30). The Excel data analysis tool for this would be the z-Test: Two-Sample for Means. Use this Excel data analysis tool only when sample size is greater than 30. The samples do not have to be the same size. On the following page is the dialogue to solve the problem.

The tool is located at:

Tools / Data Analysis / z-Test: Two-Sample for Means (in Excel 2010 -> Data tab)

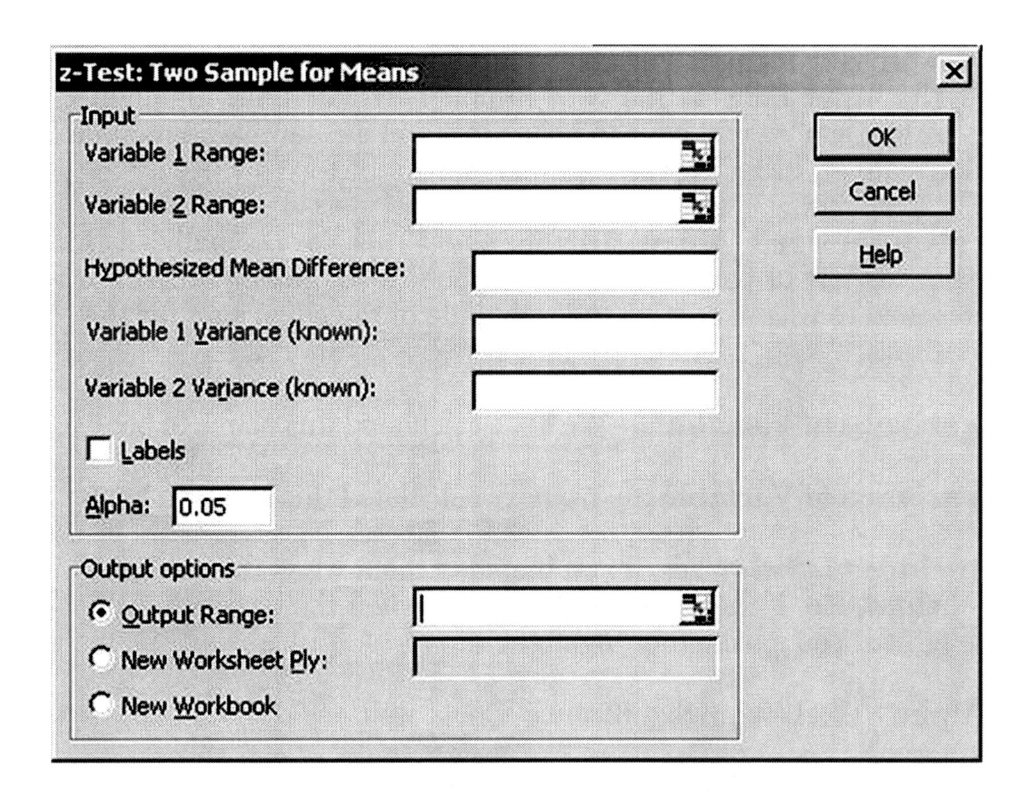

The Excel data analysis tool requires us to determine the variances of each sample first. Do that with the following Excel function typed in a cell:

= VAR (Darker Highlighted Data) = 81.3

= VAR (Lighter Highlighted Data) = 82.0

The tool's dialog box asks for the following 8 pieces of data:

1) **Variable 1 Range**: Variable 1 represents the "Before Data." In this case, the "Before Data" is the Old Supplier data column highlighted in yellow. "Before Data" would have to be arranged in a column and highlighted in this manner, including the column label on the top row.

2) **Variable 2 Range**: Variable 2 represents the "After Data." In this case, the "After Data" is the New Supplier data column highlighted in tan. "After Data" would have to be arranged in a column and highlighted in this manner, including the column label on the top row.

3) **Hypothesized Mean Difference**: This is normally 0 for a Hypothesis test of paired data. The Hypothesized Mean Difference will correspond to whatever is on the right side of the equal sign for the Null Hypothesis.

4) **Variance of Variable 1** - We have calculated this above to be: 81.3

5) **Variance of Variable 2** - We have calculated this above to be: 82.0

6) **Includes Labels?** Yes, if you included them when the columns were highlighted, No, if they were not included when the data columns were highlighted. You must either highlight and include all labels or no labels.

7) **Alpha** - The Level of Significance equals alpha, so $\alpha = 0.05$

8) **Output Range** - The output range must have at least 12 rows and 3 columns free.

Click OK and the following output will be generated:

z-Test: Two Sample for Means		
	Old Supplier	*New Supplier*
Mean	**48.74**	**53.25**
Known Variance	81.3	82
Observations	35	32
Hypothesized Mean Diffe	0	
z	-2.04	
P(Z<=z) one-tail	**0.02**	
z Critical one-tail	1.64	
P(Z<=z) two-tail	0.04	
z Critical two-tail	1.96	

Interpretation of the data is very simple: Compare the p-Value with the Level of Significance (α). If the p-Value is less than α, then the Null Hypothesis is rejected and the Alternate Hypothesis is accepted.

This is a one-tailed test so the p-value shown for the one-tailed test is the correct p-Value. In this case, the one-tailed p-Value is 0.02, which is less than the level of significance (α = 0.05) so the Null Hypothesis is rejected and the Alternate Hypothesis is accepted, which states that the batteries from the new supplier have a longer average lifetime.

The previous four Excel Hypothesis Test tools are located at:

Tools / Data Analysis (in Excel 2010 -> Data tab)

We will now discuss two Hypothesis test tools that are statistical functions. They can be found at:

Insert / Function / Statistical

The z-Test and t-Test are found here and are the Excel functions ZTEST and TTEST. Each will be discussed on the following pages:

ZTEST

The ZTEST function should not be used because it will produce an incorrect answer in certain circumstances. The ZTEST attempts to perform a Hypothesis test to determine whether a sample mean is statistically different from a population mean. The ZTEST attempts to calculate a p Value for sample mean. The user highlights a sample that is the first input of the function. ZTEST then calculates the mean of the sample and uses that information along with the other user inputs to the function (the population mean and the population or sample standard deviation) to calculate a p Value for a one-tailed, one-sample Hypothesis test.

Beware that the ZTEST only calculates a correct p Value when the sample mean is greater than the population mean.

In Excel 2003 the ZTEST calculates an incorrect p Value if the sample mean is less than the population mean. Excel 2010 does not appear to have corrected the problem.

It is highly recommended to follow instructions in the Hypothesis Testing module and NOT use the ZTEST function when performing a Hypothesis test.

Here is an explanation of why the ZTEST produces an incorrect p Value when the sample mean is less than the population mean:

The format of ZTEST is:

ZTEST (Array, μ, (σ or s))

The ZTEST function calculates the one-tailed p-Value for a one-sample Hypothesis test. The Array is the sample drawn from a population whose mean is hypothesized to be μ, The third parameter is the population standard deviation, σ, or the sample standard deviation, s, if the population standard deviation is not known.

The ZTEST then calculates a Z Score (z) by the following method:

Z Score = z = [(X_{avg} - μ) / ((σ or s) / SQRT (n))]

The Z Score, z, is the number of standard errors that X_{avg}, the sample average, is from the population mean, μ.

The ZTEST is the equivalent of the following:

ZTEST (Array, μ, (σ or s)) = 1 - NORMSDIST[(X_{avg} - μ) / ((σ or s) / SQRT (n))]

= 1 - NORMSDIST(z)

1 - NORMSDIST(z) calculates the value of the area under the Normal curve that is to the **RIGHT** of z. This would be the correct p Value if the sample mean is greater than the population mean and the p Value therefore occurs in the outer right tail.

If the sample mean is less than the population mean, the p Value will occur in the outer left tail. In this case, the p Value would be the area to the **LEFT** of z, which would be calculated as

NORMSDIST(z) , not 1 - NORMSDIST(z) that is calculated by ZTEST.

You are therefore recommended to not use ZTEST when performing a Hypothesis test. If you follow the Hypothesis testing instructions provided in the Hypothesis Testing modules in this course, you will correctly perform a Hypothesis test and obtain correct p Values in all cases.

TTEST

The TTEST function duplicates the one or two-tailed p Values that are part of the output of each of the three t-Test Data Analysis tools that are described previously. There is really no reason to use the TTEST function because the TTEST output is just one part of much more complete output that each of the t-Test Data Analysis tools provides. The t-Test Data Analysis tools described previously provide a much more complete output, but require nearly the same amount of input information as the TTEST. This will be shown below:

The TTEST has the following format:

TTEST (Array 1, Array 2, Number of Tails, Type of Test)

The arrays are highlighted by the user, just as they are for the t-Test Data Analysis tools above.

The Number of Tails can be 1 or 2.

The Type of Test has the following three choices:

1) t-Test: Paired Two Sample for Means
2) t-Test: Two:Sample Assuming Equal Variances
3) t-Test: Two-Sample Assuming Unequal Variances

It is shown as follows that the TTEST output is only a small part of the output of any of the t-Test Data Analysis tools.

t-Test: Paired Two Sample for Means
Data Analysis Tool Output

	A	B	C	D	E	F	G	H	I	J	K	L
1												
2												
3					t-Test: Paired Two Sample for Means				TTEST Outputs			
4		Array 1	Array 2									
5		3	6			Array 1	Array 2					
6		4	19		Mean	4.555555556	7.888888889					
7		5	3		Variance	6.777777778	47.11111111					
8		8	2		Observations	9	9					
9		9	14		Pearson Correlation	0.101820087						
10		1	4		Hypothesized Mean Difference	0						
11		2	5		df	8						
12		4	17		t Stat	-1.410591232						
13		5	1		P(T<=t) one-tail	0.098007892			0.098007892	=TTEST(B5:B13,C5:C13,1,1)		
14					t Critical one-tail	1.859548033						
15					P(T<=t) two-tail	0.196015785			0.196015785	=TTEST(B5:B13,C5:C13,2,1)		
16					t Critical two-tail	2.306004133						
17												

TTEST Function Output shown below is just small part of output shown above.

0.098007892	=TTEST(B5:B13,C5:C13,1,1)
0.196015785	=TTEST(B5:B13,C5:C13,2,1)

Chapter 11 - Prediction Using Regression

Basic Explanation of Regression

Multiple Regression is a statistical tool used to create predictive models. The Regression Equation - the end result of the Regression - predicts the value of an output variable (the dependent variable) based upon the values of one or more input variables (the independent variables). If there are more than one independent variable, the Regression is classified as Multiple Regression.

To begin the Regression procedure, you need completed sets of independent variables and their resulting outputs, the dependent variables. Below is an example of the data needed to calculate a Regression equation:

Dependent Variable *(Output)*	Independent Variables *(Inputs)*			
y	x_1	x_2	x_3	x_4
10	2	4	5	7
12	4	3	7	8
14	6	5	8	0
16	5	6	6	10
13	6	4	8	28

The Regression Equation

Regression Analysis will be run on the above data. The output of the Regression Analysis below is called the Regression Equation:

$\mathbf{y} = \mathbf{B_0} + (\mathbf{B_1} * \mathbf{x_1}) + (\mathbf{B_2} * \mathbf{x_2}) + (\mathbf{B_3} * \mathbf{x_3}) + (\mathbf{B_4} * \mathbf{x_4})$

$\mathbf{B_0}$, $\mathbf{B_0}$, $\mathbf{B_2}$, $\mathbf{B_3}$, and $\mathbf{B_4}$ are **Coefficients of the Regression Equation**.

This Regression Equation allows you to predict a new output (the dependent variable $\mathbf{y}$) based upon a new set of inputs (the independent variables $\mathbf{x_1}$, $\mathbf{x_2}$, $\mathbf{x_3}$, and $\mathbf{x_4}$).

Regression is for Predicting, Not Forecasting

It is very important to note that the Regression Equation is a predictive tool, not a forecasting tool. The new set of inputs must be within the range of inputs used to create the model. You should not attempt to use the Regression Equation to forecast an output based upon new inputs that are outside of the bounds of the original inputs. This is a major misuse of Regression.

Performing Multiple Regression in Excel

Instructional Video

Go to
http://www.youtube.com/watch?v=iNj6Oy2qHQw
to View a
Video From Excel Master Series
About the 4 Major
Steps to Regression in Excel
Including 2 Essential Steps
That Are Often Skipped

(Is Your Internet Connection and Sound Turned On?)

You would almost never want to do a Regression by hand. The calculations are very detailed and tedious. Furthermore, Excel contains a great data tool for performing multiple regression. It can be accessed by:

Tools / Data Analysis / Regression (in Excel 2010 -> Data tab)

Let's perform a Multiple Regression in Excel and then analyze the results after creating the Regression Equation. On the next page is a Multiple Regression Problem:

Problem: Below are the monthly rates of return of 4 stocks (Google, Yahoo, MS, and Apple) and the Tech Index. Create a Regression Equation that will predict the Tech Index return for a given month if a different set of rates of return for each company's stock is input.

	Monthly Rates of Return				
Date	Tech Index	Google	Yahoo	MS	Apple
4/1/2007	0.8799	0.7541	2.1407	-4.6296	-18.8406
5/1/2007	7.5187	14.9701	-2.5948	18.986	6.6964
6/1/2007	5.558	11.9792	7.7869	-1.7226	-3.3473
7/1/2007	1.3716	7.907	-8.5551	-0.5535	5.8442
8/1/2007	-1.6289	-5.1724	1.2474	6.679	1.9427
9/1/2007	2.4171	3.4091	0.8214	1.8261	2.1063

1st Regression Step - Graph the Data

A very important first step in performing a regression is to graph the data. Fit or lack thereof between independent and dependent variables will become apparent quickly after the data is graphed. You also want to evaluate whether the graph generally appears to be linear, or whether it might be quadratic. Excel's Regression tool only works well for graphs that are reasonably linear.

Graphing in Excel is fairly straightforward. In this case, select **Insert / Chart**. The **Standard Types** tab was then selected for this chart. Following that, the **XY (Scatter)** chart was selected. This produces a cart on a black background as seen below. You are required to select the data. A copy of the data below with yellow highlighting is what would be selected.

Monthly Rates of Return					
Date	Tech Index	Google	Yahoo	MS	Apple
4/1/2007	0.8799	0.7541	2.1407	-4.6296	-18.8406
5/1/2007	7.5187	14.9701	-2.5948	18.986	6.6964
6/1/2007	5.558	11.9792	7.7869	-1.7226	-3.3473
7/1/2007	1.3716	7.907	-8.5551	-0.5535	5.8442
8/1/2007	-1.6289	-5.1724	1.2474	6.679	1.9427
9/1/2007	2.4171	3.4091	0.8214	1.8261	2.1063

You also have to select the choice that the data is columns. The dialogue box and the resulting graph is shown on the next two pages.

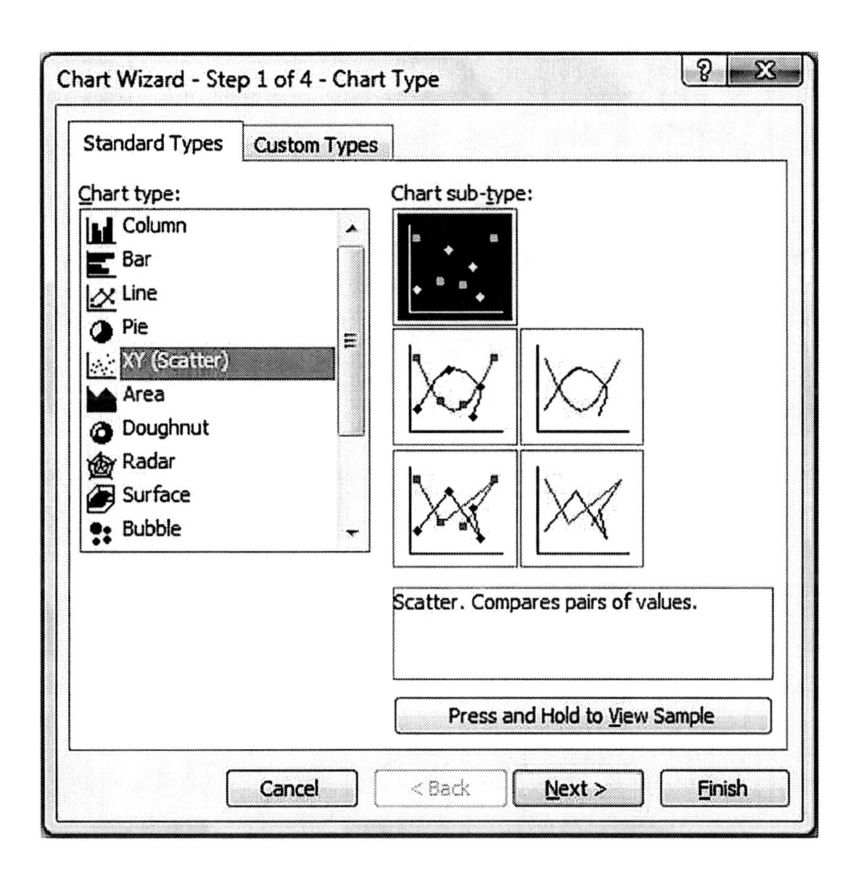
Chart Wizard - Step 1 of 4 - Chart Type
Standard Types
Custom Types
Chart type:
Column
Bar
Line
Pie
XY (Scatter)
Area
Doughnut
Radar
Surface
Bubble
Chart sub-type:
Scatter. Compares pairs of values.
Press and Hold to View Sample
Cancel
< Back
Next >
Finish

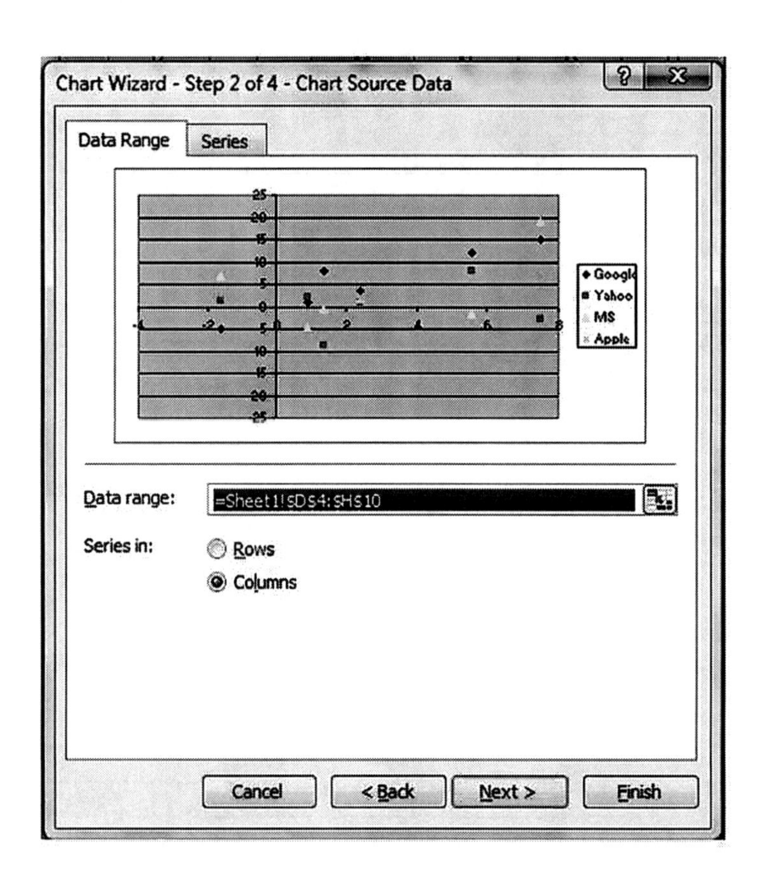

Date	Tech Index	Google	Yahoo	MS	Apple
4/1/2007	0.8799	0.7541	2.1407	-4.6296	-18.8406
5/1/2007	7.5187	14.9701	-2.5948	18.986	6.6964
6/1/2007	5.558	11.9792	7.7869	-1.7226	-3.3473
7/1/2007	1.3716	7.907	-8.5551	-0.5535	5.8442
8/1/2007	-1.6289	-5.1724	1.2474	6.679	1.9427
9/1/2007	2.4171	3.4091	0.8214	1.8261	2.1063

The highlighted data above is the data to be selected.

Type in the names of the chart and X and Y-axis names:

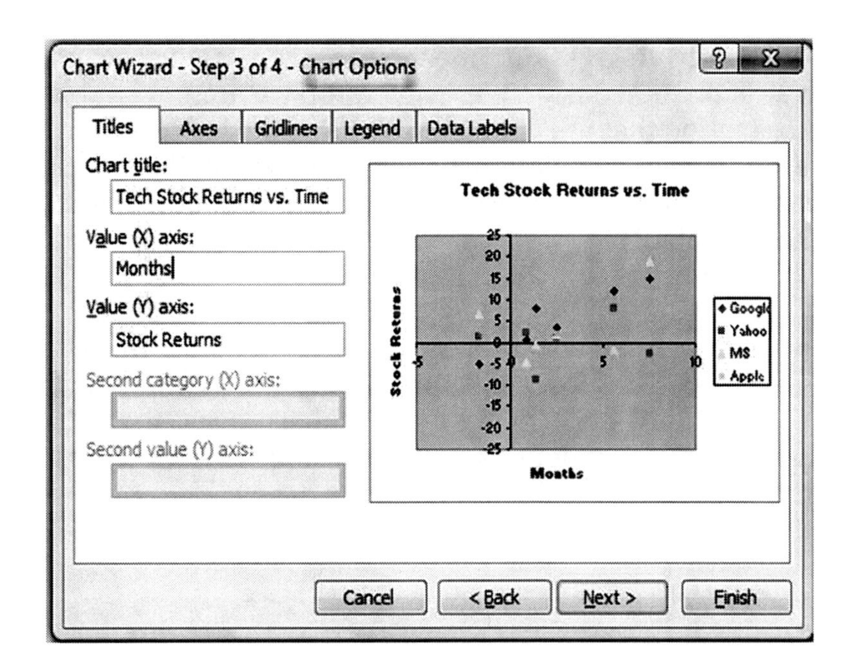

Hit "Finish" and the following chart is produced:

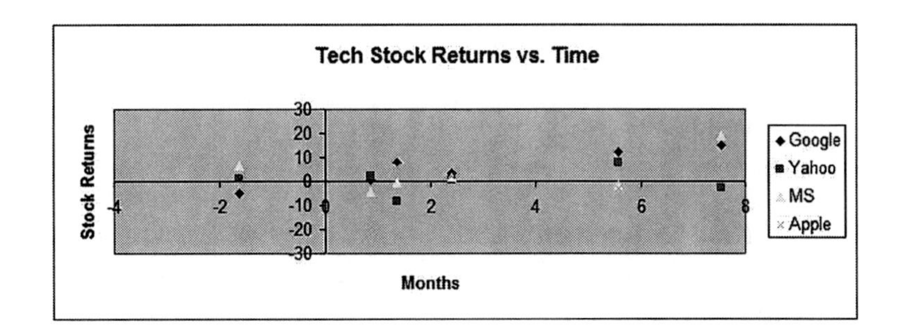

The graph of the data shows that there definitely is some relationship between the dependent variable (Tech Index) and the independent variables (Google, Yahoo, MS, and Apple).

For more detailed information about creating charts and graphs in Excel, please refer to the course module entitled "How to Graph Distributions."

2nd Regression Step - Run Correlation Analysis

The next step in the Regression process is to run a correlation analysis on all variables simultaneously. We only want to input variables in the regression equation that are good predictors of the independent variable. We will examine the correlation between the dependent variable (the output that we are trying to predict) and each of the possible inputs (the independent variables).

Correlation between two variables can take a value from anywhere between -1 and +1. The closer the correlation is to 0, the less correlated the two variables are and the less explaining power the independent variable has for the dependent variable. We want to remove any possible inputs from the regression equation if they have a low correlation with the output.

Using Excel's Correlation Tool

To access correlation analysis in Excel, follow this sequence:

Tools / Data Analysis / Correlation (in Excel 2010 -> Data tab)

The Correlation tool will require that you highlight the Data Range as is highlighted in yellow below. The data in this example is in columns and has labels in the first row. These are input settings that are requested by the Correlation tool. Specify the location of the output and press OK. You must make sure that you have sufficient area for the output. Below the correlation output is an analysis of this output.

Following is the Excel dialogue box, the data to be correlated, and the correlation analysis output:

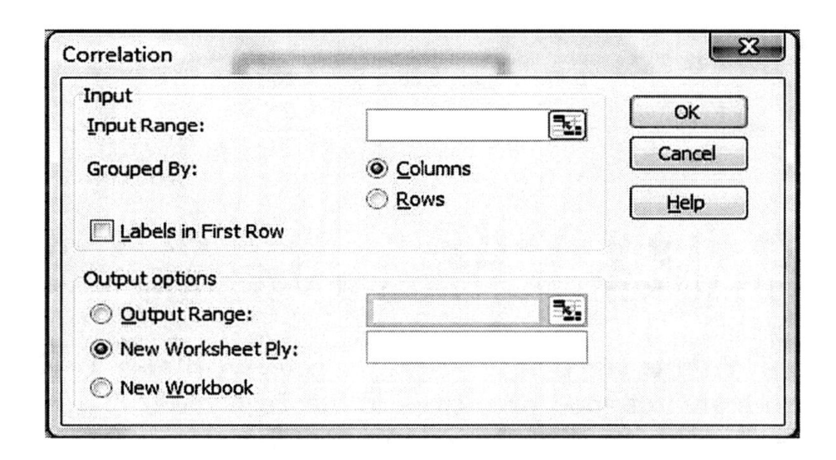

Date	Tech Index	Google	Yahoo	MS	Apple
4/1/2007	0.8799	0.7541	2.1407	-4.6296	-18.8406
5/1/2007	7.5187	14.9701	-2.5948	18.986	6.6964
6/1/2007	5.558	11.9792	7.7869	-1.7226	-3.3473
7/1/2007	1.3716	7.907	-8.5551	-0.5535	5.8442
8/1/2007	-1.6289	-5.1724	1.2474	6.679	1.9427
9/1/2007	2.4171	3.4091	0.8214	1.8261	2.1063

Correlation Analysis Output

	Tech Index	Google	Yahoo	MS	Apple
Tech Index	1				
Google	0.94	1			
Yahoo	0.13	-0.10	1		
MS	0.47	0.35	-0.26	1	
Apple	0.26	0.34	-0.50	0.63	1

Remove Input Variables that Have a Low Correlation with the Output Variable

Apple and Yahoo have low correlations with the Tech Index and therefore are not good predictors of the Tech Index. **They should be removed**. Also, if two of the independent variables above are highly correlated with each other, only one of them should be used in the Multiple Regression below. This is not the case here because none of the variables above have a high correlation with another variable. Using highly correlated variables as inputs to a Multiple Regression causes an error called Multicollinearity and should be avoided.

How To Add Input Variables to a Regression

Multiple Regressions should be built up by adding one new independent variable at a time and evaluating results. Good new independent variables noticeably raise R-Square and lower Standard Error without causing much change to Coefficients. Poor new independent variables don't change R-Square much but have unpredictable effects on Coefficients. Build regressions up one variable at a time and evaluate after adding each new variable.

R Square will always increase when a new variable is added. Adjusted R Square will only increase if the newly-added variable increases the predictive power of the Regression Equation. This is another test to determine whether a new variable adds value to the equation.

3rd Regression Step - Run Regression Analysis

Now that we decided not to include the Apple and Yahoo returns in the regression analysis, let's do this Regression Analysis. Go to the Tool Dropdown menu and follow this sequence;

Tools / Data Analysis / Regression (in Excel 2010 -> Data tab)

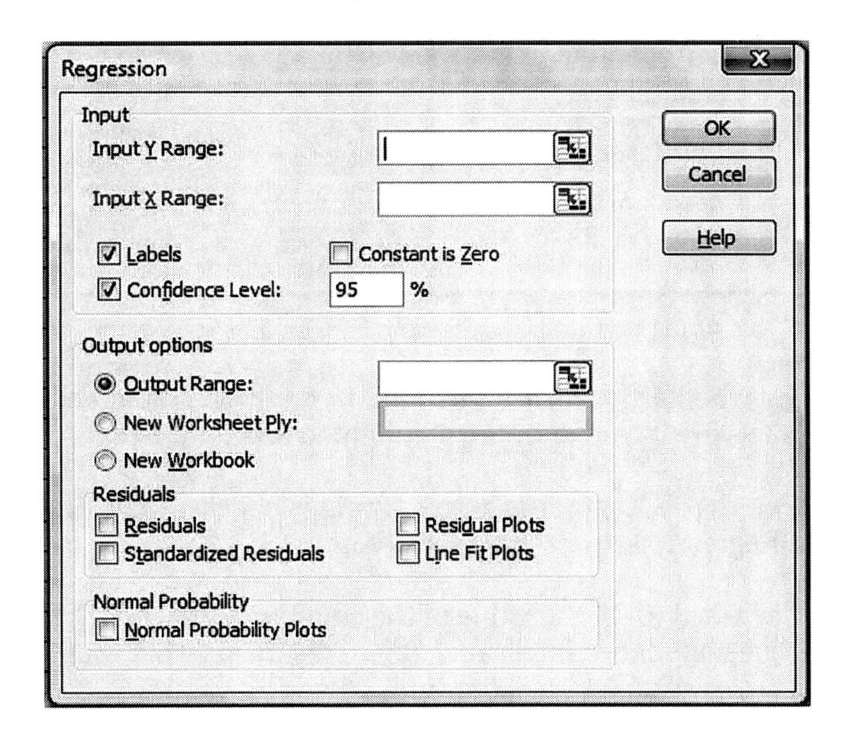

Highlight the x Range. Here, all independent variables and their labels would be highlighted. The x Range for this problem is highlighted darker below. The variables Google and MS are the x (independent) variables.

Highlight the y Range. Here, the one dependent variable and its label would be highlighted. The y Range for this problem is highlighted lighter below. Tech Index is the Y (dependent) variable.

Tech Index	Google	MS
0.8799	0.7541	-4.6296
7.5187	14.9701	18.986
5.558	11.9792	-1.7226
1.3716	7.907	-0.5535
-1.6289	-5.1724	6.679
2.4171	3.4091	1.8261

Next specify whether labels are included in the highlighted cells. In this case, the labels are included in the highlighted area.

Do not check that "Constant is zero." If this is checked, the first term of the output Regression, $\mathbf{B_O}$, will be set to zero.

Next you are asked to set the value of the confidence interval. The default setting for the confidence interval is 95%. This means that there is a 95% chance that regression analysis output is correct.

Now set the location of the upper left corner of the output. Make sure that there is enough room for the entire output in the location that you designate. Following is the output of the Regression Analysis.

Regression Analysis Output

SUMMARY OUTPUT

Regression Statistics	
Multiple R	0.951
R Square	0.904
Adjusted R Square	0.840
Standard Error	1.331
Observations	6

ANOVA

	df	SS	MS	F	Significance F
Regression	2	49.970	24.985	14.106	0.030
Residual	3	5.314	1.771		
Total	5	55.284			

	Coefficients	Standard Error	t Stat	P-value	Lower 95%	Upper 95%
Intercept	0.251	0.711	0.354	0.747	-2.011	2.514
Google	0.393	0.085	4.616	0.019	0.122	0.664
MS	0.063	0.075	0.843	0.461	-0.175	0.300

Regression Equation Tech Index = (0.251) + (0.393)*(Google) + (0.063)*(MS)

4th Regression Step - Analyze the Output

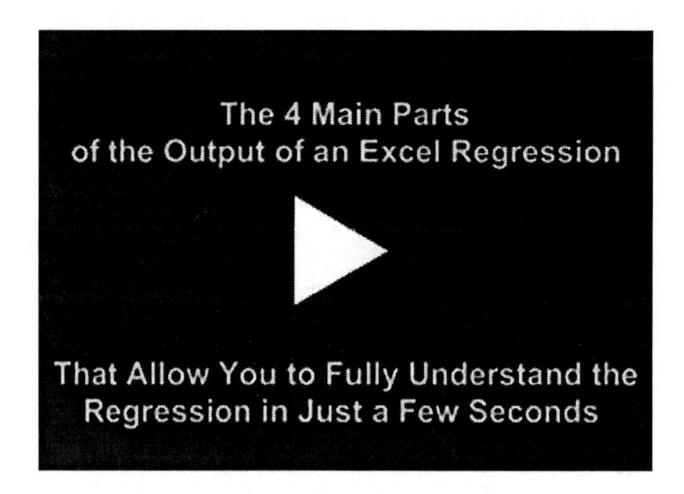

Instructional Video

Go to
http://www.youtube.com/watch?v=ECXeUj8I6w8
to View a
Video From Excel Master Series
About How To
Quickly Read and Understand
the 4 Main Parts of Excel's
Regression Output

(Is Your Internet Connection and Sound Turned On?)

The Regression Equation

The Regression Output that we are most interested in is the Regression Equation. The equation has the following form:

$\mathbf{y} = \mathbf{B_o} + (\mathbf{B_1} * \mathbf{x_1}) + (\mathbf{B_2} * \mathbf{x2})$

$\mathbf{B_o}$, $\mathbf{B_1}$, and $\mathbf{B_2}$ are **Coefficients of the Regression Equation**.

The cells containing coefficients,Bo, B1, and B2 are highlighted in light blue in the output. The Regression Equation is therefore as follows:

Tech Index = (**0.251**) + (**0.393**)*(**Google**) + (**0.063**)*(**MS**)

Using the Regression Equation to Predict an Output

Problem: Predict the Tech Index if Google = 7 and MS = 4.

Tech Index = (0.251) + (0.393)*(Google) + (0.063)*(MS)

$\text{Tech Index}_{\mathbf{predicted}} = (0.251) + (0.393)*(7) + (0.063)*(4) = 3.25$

The Confidence Interval of the Output Variable

The Confidence Interval was set at 95%. This is the default setting. It could have been set to any desired confidence level.

The 95% Confidence Interval is the interval in which the output variable, Tech Index, should fall with 95% certainty.

The 95% Confidence Interval = $\text{Tech Index}_{\mathbf{predicted}}$ +/- $\text{Z Score}_{\mathbf{95\%}}$ * (Standard Error)

Z Score Calculation

Level of Confidence = 95% = 1 - α

Level of Significance = α = 0.05

Z $Score_{95\%,2\text{-tailed}}$

= NORMSINV(1 - α/2)

= NORMSINV (1 - 0.025)

= NORMSINV (0.975) = 1.96

The cell containing the Overall Standard Error for the Regression Equation is highlighted in orange in the output. Its value = 1.33

The 95% Confidence Interval = Tech $Index_{predicted}$ +/- Z $Score_{95\%}$ * Standard Error

The 95% Confidence Interval = 3.25 +/- (1.96) * (1.33) = 0.64 to 5.86

This means that there is a 95% chance the actual Tech Index return will fall within 0.64 and 5.86 for inputs Google = 7 and MS = 4.

R Square

After the Regression Coefficients, the 2nd most important output is the R Square. R Square explains what percentage of total variance in the output is explained by the variance of the inputs. R^2 (R Square) is one measure of correlation between the input and the output, very similar to r as the correlation coefficient between 2 variables. R^2 tells how well the regression line approximates the real data.

Here, R^2 (R Square) is 0.904. This means that 90.4% of the variation of the Tech Index can be explained by variation within the input variables. Ideally, you would like to see R Square to be at least 0.50 (50% of the variance in the output is explained by variance of the inputs.

Adjusted R Square

Closely related to R Square is the Adjusted R Square. This value is created by adjusting the R Square for the number of explanatory terms in the model. Unlike R^2, the Adjusted R^2 increases only if the new term improves the model more than would be expected by chance. The adjusted R^2 can be negative, and will always be less than or equal to R^2.

Adjusted R Square is quoted much more often than R Square when describing the accuracy of a regression. One reason that the Adjusted R Square is more conservative that R Square is because it is always less than R Square. Another reason is that when new input variables are added to the regression equation, R Square will always increase. Adjusted R Square, on the other hand, only increases if the new input variables increase the percentage of variance that is explained by the input variables. In other words, Adjusted R Square increases only when a newly-added variable increases the predictability of the output (the dependent variable).

Adjusted R Square is only a better measure if R Square was calculated based upon a sample. If R Square was calculated using the entire population, then Adjusted R Square has no advantage over the Adjusted R Square.

Adjusted R Square is simply R Square adjusted for the number of explanatory terms.

In this case, the Adjusted R Square is 0.839. Adjusted R Square is always less than R Square (0.903 here) and is therefore a more conservative estimate of total variance of the output variable that is explained by the input variables.

One more note on Adjusted R Square - It is best to build models starting with one input variable and then add input variables one at a time to see how much each new variable increases Adjusted R Square.

F Statistic

The Excel regression function runs an F test to determine the validity of the overall regression output. The F test answers the question of what is the likelihood that the result could have been obtained merely by chance. If the likelihood of a chance output is very small, the validity of the output is confirmed.

The Significance of F - cells highlighted in purple - provides the probability that the overall result could have been obtained by chance. If this case, the Significance of F is 0.030. That says that there is only a 3.0% chance that the output produced by the regression function was obtained by chance. This is a strong result. We can feel confident that this regression has produced a valid result. For those familiar with ANOVA testing, the Significance of F is normally referred to as the p value of the F test. For more detailed information on ANOVA testing, refer to that module of this course.

As stated in the output, the F test is part of an ANOVA (Analysis of Variance) test. An ANOVA test analyzes all of the sources of variance and determines the likelihood that the amounts of variance from each source occurred by chance or not. The p value (called here by Excel as the Significance of F) provides the overall likelihood that the amount of variance attributed to each source of variance is correct. In other words, the p value provides the likelihood that the result occurred by chance.

In this case, the low p value (Significance of F) confirms the R Square value of 0.903 which states that 90.3% of the total variance can be explained by and is derived from the independent variables (the inputs). The remainder of the variance, only 9.7% of the total variance, is unexplained. The ANOVA test confirms the validity of this result through the low p value (Significance of F).

ANOVA Calculations of the Regression Output

Calculations for the ANOVA test performed by Excel's regression function are as follows:

Overall variance of the output can be divided into two sources: Explained Variance and Unexplained Variance. Explained Variance is labeled here as Regression. This consists of all variance that can be explained by and attributed to the independent input variables. Unexplained Variance is labeled here as Residual. As with a standard ANOVA test, each type of variance has the following three calculations associated with it. These are Degrees of Freedom (df), Sum of Squares (SS), and Mean Square (MS).

The three ANOVA calculations (Degrees of Freedom, Sum of Squares, and Mean Squares) associated with the Regression inputs are in grey-highlighted cells. The ANOVA calculations associated with Unexplained Variance are light-purple highlighted cells.

For a detailed description of the ANOVA calculation, please refer to the ANOVA module of this course. Within that module is a complete hand calculation of an entire ANOVA function.

Here are some of the important highlights of the ANOVA calculation that appears in the Excel Regression output.

There are two degrees of freedom associated with the Explained Variance and three degrees of freedom associated with Unexplained Variance. These combine for a total of five degrees of freedom.

Mean Square (**MS**) for both the Explained Variance (Regression) and the Unexplained Variance equals its Sum of Squares (**SS**) divided by its degrees of freedom (**df**) as follows:

$MS_{Explained} = SS_{Explained} / df_{Explained} = 49.970 / 2 = \mathbf{24.985}$

$MS_{Unexplained} = SS_{Unexplained} / df_{Unexplained} = 5.314 / 3 = \mathbf{1.771}$

F Statistic = **Explained Variance / Unexplained Variance**

$= MS_{Explained} / MS_{Unexplained} = \mathbf{24.985 / 1.771 = 14.106}$

Significance of F (often referred to as the p statistic)

= FDIST(**14.106**,2,3)

= 0.030

As stated previously, the Significance of F indicates the likelihood that the Regression output could be obtained by chance. In this case, there is only a 3.0% chance that the Regression output could have occurred merely by chance. In other words, this small F Statistic strongly confirms the validity of the Regression output.

To sum up the meaning of the F Statistic, the F Test here calculates the ratio of Explained Variance over Unexplained Variance. If this ratio Is large enough, it unlikely that this result was obtained by chance. This large ratio confirms that the prediction made by the regression equation is an improvement over a prediction made by chance.

P Values of Regression Coefficients and Intercept

The P values associated with each Regression Coefficient and the Y-Intercept of the Regression output provide the degree of validity of each. The lower the P value, the greater the likelihood of validity of that coefficient or intercept.

The P value of the Google coefficient is 0.019. This is the lowest P value, which indicates that this coefficient has the highest likelihood of being valid. We can conclude from its P value of 0.019 that this coefficient differs significantly from zero with a 5% maximum possibility of error (level of significance, $\alpha = 0.05$), but not at the 1% level of significance (no more than 1% chance that the coefficient is different from zero). This lowest P value implies that the Google return is a much better predictor of the Tech Index return than the MS return, which has a much higher P value.

The P value for each coefficient and intercept is conveniently calculated by the Excel Regression function. These calculations are fairly involved and will not be performed here.

Some of the other more basic calculations in the process of determining the P value are shown here:

The t Statistic for each coefficient and intercept equals that coefficient or intercept divided by its Standard Error.

For example:

$\text{t Stat}_{\text{Google}} = \text{Coefficient}_{\text{Google}} / \text{Standard Error}_{\text{Google}}$
$= 0.393 / 0.085 = 4.616$

The 95% confidence interval for each coefficient and intercept is provided. This range can be evaluated using the Google return as an example. There is 95% certainty that the true Regression Coefficient for the Google return is between 0.122 and 0.664.

Regression Using Dummy Variables

"Dummy Variables" are used in Regressions to represent variables that have only two discrete states. Dummy variables used in the Regression Equation can take only the values 1 or 0.

Instructional Video

Go to
http://www.youtube.com/watch?v=EMbiGPGlBEM
to View a
Video From Excel Master Series
About How To Use
Dummy Variable Regression
in Excel To Perform
Conjoint Analysis

(Is Your Internet Connection and Sound Turned On?)

Dummy Variables are used to describe whether an object has a certain attribute or not. If the object has the attribute in question, the Dummy Variable assumes the value of 1. If the object does not have the attribute, the Dummy variable assumes the value of 0.

An example of such an attribute would be color. For example, the object described is either blue or it is not. If the object is blue, the Dummy Variable has the value of 1. If the object is not blue, the Dummy Variable has the value of 0.

Creating Dummy Variables for Attributes with Multiple Choices

If an attribute can have more than two choices, one unique Dummy Variable must be created for each choice. For example, if an object has multiple choices of color, such as red, yellow, or blue, a separate Dummy Variable must be created for each color. The Dummy Variable for the color can be one of three colors (red, yellow, and green for example), red will have the value of 1 if the object is red. If not, this Dummy Variable is 0. A similar Dummy Variable must be created for the color yellow and for the color green.

One problem with using Dummy Variables to represent more than one state of an attribute is that the value of the final Dummy Variable can be predicted if you know the value of the other Dummy Variable or Variables. The following is an example:

If an object can come in the three colors red, yellow, or green and you know the value of two of the Dummy Variables (for example, you know the value of the Red Dummy Variable and the Yellow Dummy Variable), then you can determine the value of the final Green Dummy Variable. If the value of either of the Red or Yellow Dummy Variables is 1, then the value of the Green Dummy Variable must be 0. If the values of both the Red and Yellow Dummy Variables are 0, the Green Dummy Variable must be 1.

Removing a Dummy Variable to Prevent Collinearity

The error described above is called Collinearity. Collinearity occurs if any of the independent variables can be used to predict the value of any of the other independent variables. This problem can be solved by removing one of the Dummy Variables from any set of Dummy Variables that describes the same attribute.

In the current example, the problem of Collinearity will be solved if either the Red, Yellow, or Green Dummy Variable is removed from the input of the Regression Equation. We will see in the example below that removing one variable in each set of attributes does not have any detrimental effect upon the completeness or accuracy of the output.

Below is an example of a Regression that requires multiple Dummy Variables. This example also illustrates how a well-known marketing technique called Conjoint Analysis is performed.

Conjoint Analysis Done With Regression Using Dummy Variables

Conjoint Analysis is used by marketers to tell which attributes of a product are most important to a consumer and also to quantify the degree of importance of an attribute. The degree of importance that a consumer places on a product attribute is called the "utility" of that attribute.

For example, a product might simultaneously come in three brands, two colors, and three levels of price. Each color, brand, and price level will have its own utility calculated during the conjoint analysis. Conjoint Analysis is done using Multiple Regression. Each product attribute variation will be assigned as one of the independent variable inputs to the Multiple Regression equation.

In the above example, the color red will be represented by one independent variable while the color blue will be represented by another independent variable. The resulting regression equation assigns a coefficient to each independent variable. These coefficients are the utilities of each of the attributes. The more positive an individual coefficient is, the more highly valued is the associated product attribute. The coefficients of the Regression Equation can be interpreted as the utilities of the variables.

On the following pages are the steps to performing Conjoint Analysis in Excel:

1st Conjoint Step - List Product Attributes

In this conjoint exercise, we are going to determine the utilities of eight product attributes. They are listed as follows:

Brand	Color	Price
A	Red	$50
B	Blue	$100
C		$150

2nd Conjoint Step - List All Attribute Combinations

There are 18 possible combinations of these attributes: (3 brands x 2 colors x 3 prices). The consumer rates each combination on a scale of 0 to 10 (10 being the best). The consumer test results are modified for the regression equation and then run through the regression. The resulting regression analysis calculates a coefficient for each independent variable as part of the regression output equation. Each coefficient is the measure of value that the consumer places on the product attribute.

The following chart provides the choices that the consumer had to analyze. The consumer was provided with 18 separate cards. Each card contained one of the 18 possible variations of product attributes. The consumer had to rate their overall preference of each combination of attributes on a scale of 1 to 10.

Card	Brand	Color	Price
1	A	Red	50
2	A	Red	100
3	A	Red	150
4	A	Blue	50
5	A	Blue	100
6	A	Blue	150
7	B	Red	50
8	B	Red	100
9	B	Red	150
10	B	Blue	50
11	B	Blue	100
12	B	Blue	150
13	C	Red	50
14	C	Red	100
15	C	Red	150
16	C	Blue	50
17	C	Blue	100
18	C	Blue	150

3rd Conjoint Step - Conduct Consumer Survey

Have the consumers rank each combination on a scale of 1 (worst) to 10 (best). The chart below shows the consumer's stated preference for each combination of attributes. Non-numerical attributes were assigned numbers. Brand A and Red are shown as 1's in their respective columns. Brand B and Blue were shown as 2's in their respective columns. Brand C was assigned a 3 in its respective column. Normally each possible attribute combination is placed on a separate card. The consumer ranks the cards in order of preference and then assigns a level of preference from 1 to 10 on each card. Below is one summary of a consumer's choices:

Card	Brand	Color	Price	Preference
1	1	1	50	5
2	1	1	100	5
3	1	1	150	0
4	1	2	50	8
5	1	2	100	5
6	1	2	150	2
7	2	1	50	7
8	2	1	100	5
9	2	1	150	3
10	2	2	50	9
11	2	2	100	6
12	2	2	150	5
13	3	1	50	10
14	3	1	100	7
15	3	1	150	5
16	3	2	50	9
17	3	2	100	7
18	3	2	150	8

4th Conjoint Step - Create Dummy Variables for Attributes

Dummy Variables must be created for each level of each product attribute. A Dummy Variable is assigned the value of 1 if that attribute level of its represented attribute is selected by the consumer. Otherwise the Dummy Variable is assigned the value of 0. Below shows the consumer's ranking above converted to Dummy Variables:

	Dummy Variable	Dummy Variable	Dummy Variable	Dummy Variable	Dummy Variable	Dummy Variable	Dummy Variable	Dummy Variable	
Card	A	B	C	Red	Blue	$50	100	150	Preference
1	1	0	0	1	0	1	0	0	5
2	1	0	0	1	0	0	1	0	5
3	1	0	0	1	0	0	0	1	0
4	1	0	0	0	1	1	0	0	8
5	1	0	0	0	1	0	1	0	5
6	1	0	0	0	1	0	0	1	2
7	0	1	0	1	0	1	0	0	7
8	0	1	0	1	0	0	1	0	5
9	0	1	0	1	0	0	0	1	3
10	0	1	0	0	1	1	0	0	9
11	0	1	0	0	1	0	1	0	6
12	0	1	0	0	1	0	0	1	5
13	0	0	1	1	0	1	0	0	10
14	0	0	1	1	0	0	1	0	7
15	0	0	1	1	0	0	0	1	5
16	0	0	1	0	1	1	0	0	9
17	0	0	1	0	1	0	1	0	7
18	0	0	1	0	1	0	0	1	8

5th Conjoint Step - Remove 1 Dummy Variable from Each Set of Attributes

One problem must be corrected before this data can be submitted for Regression Analysis. Independent variables or combinations of independent variables should not be able to predict each other. Using independent variables that are highly correlated with each other (either positively or negatively) produces a regression error known as Collinearity.

For example, if the color is either red or blue, knowing the state of one of the color (if the state of Blue = 1, the state of Red must = 0), we know the state of the other color.

This error condition also occurs when there are 3 variables. If you know the states of 2, then you know the state of the remaining one.

These error conditions are solved by removing one column of data from each type of variation. Information about Brand A, Red, and Price level $50 were removed.

We will see later that this has no effect on the accuracy of the Regression output.

Card	B	C	Blue	$100	$150	Preference
1	0	0	0	0	0	5
2	0	0	0	1	0	5
3	0	0	0	0	1	0
4	0	0	1	0	0	8
5	0	0	1	1	0	5
6	0	0	1	0	1	2
7	1	0	0	0	0	7
8	1	0	0	1	0	5
9	1	0	0	0	1	3
10	1	0	1	0	0	9
11	1	0	1	1	0	6
12	1	0	1	0	1	5
13	0	1	0	0	0	10
14	0	1	0	1	0	7
15	0	1	0	0	1	5
16	0	1	1	0	0	9
17	0	1	1	1	0	7
18	0	1	1	0	1	8

6th Conjoint Step - Run Regression Analysis

Here is the output of the Regression analysis:

SUMMARY OUTPUT

Regression Statistics	
Multiple R	0.93
R Square	0.87
Adjusted R Square	0.81
Standard Error	1.14
Observations	17

ANOVA

	df	*SS*	*MS*	*F*	*Significance F*
Regression	5	96.61	19.32	14.83	0.000143011
Residual	11	14.33	1.30		
Total	16	110.94			

	Coefficients	*Standard Error*	*t Stat*	*P-value*	*Lower 95%*	*Upper 95%*
Intercept	5.92	0.81	7.33	0.00001	4.14	7.69
Brand B	1.51	0.70	2.17	0.05314	-0.02	3.05
Brand C	3.35	0.70	4.79	0.00056	1.81	4.89
Blue	1.23	0.56	2.20	0.05016	0.00	2.46
$100	-2.32	0.70	-3.32	0.00685	-3.86	-0.78
$150	-4.32	0.70	-6.18	0.00007	-5.86	-2.78

Regression Equation   Combination Preference = 5.912 + (1.51)*(Brand B) +

+ (3.35)*(Brand C) + (1.23)*(Blue) + (-2.32)*($100) + (-4.32)*($150)

7th Conjoint Step - Analyze the Output

The regression appears to be a good one because Adjusted R Squared is high (close to 1). Adjusted R Square equals Explained Variance divided by Unexplained Variance and then adjusted for the The number of explanatory terms. Here, Adjusted R Square is 8.12.

Each of the variables has a low p-Value and is therefore a significant predictor.

The absolute value of the coefficients indicates the effect that each has on the consumer's overall liking of product. For example, Brand C (coefficient = 3.347) produced the highest positive influence while the $150 price (coefficient = -4.319) reduces consumer liking the most.

Showing That Removing Dummy Variables Did Not Affect Output

Removing information about Brand A, Red, and Price level $50 did not hurt the output accuracy. These product attributes could still be considered to be part of the Regression equation, but with coefficients of 0 in the regression equation.

The coefficients attached to each of the product attributes simply show the consumer's utility for that attribute. The utilities for each attribute are relative to each other.

Using the results from the 13th card, we can demonstrate that the removal of the three Dummy Variables did not affect the outcome. This is shown in the diagrams on the next page.

Card	B	C	Blue	$100	$150	Preference
1	0	0	0	0	0	5
2	0	0	0	1	0	5
3	0	0	0	0	1	0
4	0	0	1	0	0	8
5	0	0	1	1	0	5
6	0	0	1	0	1	2
7	1	0	0	0	0	7
8	1	0	0	1	0	5
9	1	0	0	0	1	3
10	1	0	1	0	0	9
11	1	0	1	1	0	6
12	1	0	1	0	1	5
13	0	1	0	0	0	**10**
14	0	1	0	1	0	7
15	0	1	0	0	1	5
16	0	1	1	0	0	9
17	0	1	1	1	0	7
18	0	1	1	0	1	8

Regression Equation Combination Preference = 5.912 + (1.51)*(Brand B) +

+ (3.35)*(Brand C) + (1.23)*(Blue) + (-2.32)*($100) + (-4.32)*($150)

which is equivalent to:

Combination Preference = 5.912 + **(0)*(Brand A)** + (1.51)*(Brand B) + (3.35)*(Brand C) + **(0)*(Red)** + (1.23)*(Blue) + **(0)*($50)** + (-2.32)*($100) + (-4.32)*($150)

Each Dummy Variable that was removed is assigned the value of 0. The removed Dummy Variables are: a) Price level $50, b) Color red, and c) Brand A. Each of these is assigned a coefficient of 0.

For example, Price level $50 has the highest preference with a utility of 0 while Price level $150 has the lowest utility of -4.32. Blue has a utility of 1.23, which is that much higher than the utility of red, which was 0. Brand C was the most liked brand with a utility of 3.347 with Brand A is liked the least with a utility of 0.
The resulting Regression Equation still does a good job of predicting overall preference. For example, the consumer rated the combination of attributes on card 13 with a 10.

Combination Preference = 5.912 + **(0)*(Brand A)** + (1.51)*(Brand B) + (3.35)*(Brand C) + **(0)*(Red)** + (1.23)*(Blue) + **(0)*($50)** + (-2.32)*($100) + (-4.32)*($150)

Card 13 Combination Preference = 5.912 + **(0)*(Brand A)** + (1.51)*(0) + (3.35)*(1) + **(0)*(Red)** + (1.23)*(0) + **(0)*($50)** + (-2.32)*(0) + (-4.32)*(0)

Card 13 Combination Preference = 5.912 + **0** + 0 + 3.35 + **0** + 0 + **0** + 0 + 0 = 9.26

Here the predicted Combination Preference for card 13 attribute combination is equals 9.26, which is very close to the consumer's rating of 10.

Hand Calculation of Regression Problems

Go To
http://excelmasterseries.com/Excel_Statistical_Master/Regression.php

To View How To Solve Regression Problems By Hand (No Excel)

(Is Your Internet Connection Turned On ?)

You'll Quickly See Why You Always Want To Use Excel To Solve Statistical Problems !

Chapter 12 - Independence Tests and ANOVA

Basic Explanation of ANOVA

ANOVA, Analysis of Variance, is a test to determine if three or more different methods or treatments have the same effect on a population. For example, ANOVA testing might be used to determine if three different teaching methods produce the same test scores with a group of students. The measured output must be some type of group average such as average test score per group or average sales per group. ANOVA testing might also be used to determine if different combinations of product pricing and promotion have different effects in different markets.

In summary, ANOVA testing is used to judge whether three or more groups have the same mean (for example, same test scores) after each group has had a **different treatment** applied to it (for example, a different teaching method applied to each group).

ANOVA Tests the Null Hypothesis - That Nothing Is Different Between Groups

The basic test of ANOVA is the Null Hypothesis that the different methods had no effect on the outcome that is being measured. Using the teaching method example, the Null Hypothesis in this case would be that the different teaching methods had no effect on the average test scores of student groups to which different treatments (teaching methods) were applied.

The Null Hypotheses, which is being tested, states that the average test score from each of the three groups or populations to which a different teaching method was applied should be the same.

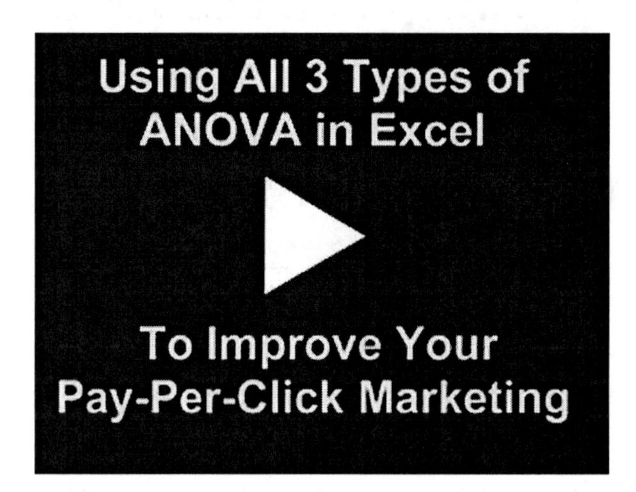

Instructional Video

Go to
http://www.youtube.com/watch?v=1nddyCJLAOc
to View a
Video From Excel Master Series
About Doing
All 3 Types of ANOVA in Excel

(Is Your Internet Connection and Sound Turned On?)

Overview of ANOVA in Excel

The hand calculations for ANOVA are very tedious. Excel has a built-in ANOVA function that does a great job. The problems below are completed using Excel's built-in ANOVA function. The simplest example, known as Single Factor ANOVA, is completed using both Excel's built-in ANOVA function and also hand-calculated at the very end of this course module.

The ANOVA tests in Excel can be accessed by the drop-down menu:

Tools / Data Analysis (in Excel 2010 -> Data tab)

Inside Excel's Data Analysis menu, there are three types of ANOVA analysis available:

1) Single Factor ANOVA
2) Two-Factor ANOVA Without Replication
3) Two-Factor ANOVA with Replication

Each ANOVA test type is explained below:

Single Factor ANOVA

Single Factor ANOVA tests the effect of just one factor, in this case, the teaching method, on the measured outputs. The measured outputs are the mean test scores for the groups that had the different teaching methods applied to them. The Null Hypothesis for this one factor states that varying that factor has no effect on the outcome.

Two-Factor ANOVA Without Replication

Two-Factor ANOVA Without Replication - Allows testing of the original factor plus one other factor. For example, in addition to testing teaching methods, you could also test an additional factor, such as whether differences in teaching ability caused additional variation in the outcome of test average scores. Each factor has a Null Hypothesis which states that varying that factor had no effect on the outcome.

Two-Factor ANOVA With Replication

Two-Factor ANOVA With Replication allows for testing both factors as above. This method also allows us to test the effect of interaction between the factors upon the measured outcome. The test is replicated in two places. This allows for analysis of whether the interaction between the two factors has an effect on the measured outcome. The Null Hypothesis for this interaction test states that varying the interaction between the two factors has no effect on the measured outcome. Each of the other two factors being tested also has its own Null Hypothesis.

Instructional Video

Go to
http://www.youtube.com/watch?v=K4ZVnsE17IE
to View a
Video From Excel Master Series
About Doing
ANOVA in Excel
and Doing It By Hand -
Both Ways Are Shown Completely

(Is Your Internet Connection and Sound Turned On?)

Examples of each type of ANOVA performed with Excel follow:

ANOVA: Single Factor Analysis

The Hand Calculation of this problem's ANOVA is performed at the end of this course module.

Problem 1: Three Sales Closing Methods and Single Factor ANOVA

Problem: Three different sale closing methods were used. Three groups of four salespeople were randomly chosen. Each group was instructed to use only one of the closing methods for all of their sales. Sales totals of each salesperson over the next two weeks were collected. Determine with a 95% level of certainty whether there is a difference in the effectiveness of the closing methods. Following are sales results for all 12 salespeople on the next page:

Sales Group 1	Closing Method 1
Salesperson 1	16
Salesperson 2	21
Salesperson 3	18
Salesperson 4	13

Sales Group 2	Closing Method 2
Salesperson 5	19
Salesperson 6	20
Salesperson 7	21
Salesperson 8	20

Sales Group 3	Closing Method 3
Salesperson 9	24
Salesperson 10	21
Salesperson 11	22
Salesperson 12	25

The above information must be arranged as follows for input into Excel:

Sales Group ---->	Group 1	Group 2	Group 3
Treatment ---->	Method 1	Method 2	Method 3
Salesperson A	16	19	24
Salesperson B	21	20	21
Salesperson C	18	21	22
Salesperson D	13	20	25

Problem Solving Steps

This is a **Single Factor ANOVA** test because we are testing only whether different variations of a single factor (Closing Method) have an effect on measured outcome (sales of each salesperson) using a different method. Nothing else is entered into the test that might have an effect on the measured outcome. **The abilities of all salespeople are assumed to be similar. Each individual salesperson will use only one of the closing methods.**

The **Null Hypothesis** for this test states that the closing methods used will have no effect on the measured output (sales).

Level of Certainty = 95% = 1 – α

Level of Significance = Alpha = α = 0.05

Excel Instructions:

Tools / Data Analysis / Anova: Single Factor (in Excel 2010 -> Data tab)

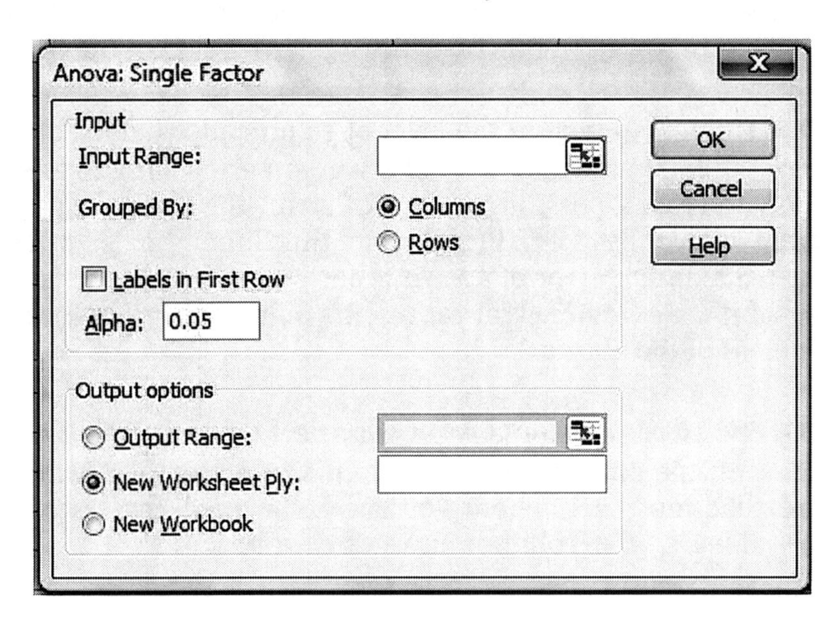

You will be required to provide the following 5 inputs:

1) Input Range: Highlight the data and labels as is highlighted in yellow previously. Note that your data must be initially arranged in columns as above or similarly in rows. Note that only the Method labels and not the Salesperson labels are highlighted because Single Factor ANOVA tests the effects of only one factor, the training method. If you also wanted to test the additional factor of whether the individual abilities of each salesperson were also a factor in the outcome, the salesperson labels would be highlighted and Two-Factor ANOVA would be used. The next example in this module performs this test.

2) Grouped By: Select Columns or Rows, depending on how your data is arranged. In this case, the data is arranged in columns

3) Data Labeling: Select whether your data is labeled in the first column (if the data is grouped in rows) or labeled in the first row (if your data is grouped in rows). In this case, the data is labeled in the first row.

4) Alpha: You will type in the level of significance here. The level of significance (alpha) is the max level of error you are willing to tolerate. If you can only tolerate a possibility of 1% of error, the level of significance (alpha) here would be 0.01. Prefilled in this input box is 0.05. This represents a maximum error of 5%. Another way of looking at this would be to say that 5% maximum error represents being at least 95% certain of the correctness of the output.

5) Output: Here you would enter (or select) the upper-left corner of where you want the output printed on your worksheet. You will need to make sure that the place where you have specified the output to be printed has at least 7 free columns and 15 free rows.

Hit "OK" and Excel will perform the Single Factor ANOVA analysis and output the following result where you have specified:

Anova: Single Factor

SUMMARY

Groups	Count	Sum	Average	Variance
Method 1	4	68	17	11.33333333
Method 2	4	80	20	0.666666667
Method 3	4	92	23	3.333333333

ANOVA

Source of Variation	SS	df	MS	F	P-value	F crit
Between Groups	72	2	36	7.043478261	0.014419201	4.256494729
Within Groups	46	9	5.111111111			
Total	118	11				

P-value = 0.014419201

Alpha = 0.05

Analyze the Output

The P-value

Without going into too much statistical detail, the cell in the output entitled P-value will tell you whether the three different closing methods have an effect on the measured output (sales of salespeople using the different training methods). Highlighted on the Excel output is the P-value for Between Groups variation.

General Rule:

If P-value < alpha ----> The different treatments affected the output

If P-value > alpha ----> The different treatments did not affect the output

If this P-value (0.01441920) is less than the level of significance (Alpha = 0.05) that was specified as one of the Excel inputs, we can say with 95% certainty (having a maximum chance of error of 5%) that the closing methods produced different effects on the measured output (sales)

In this case, the P-Value for between Group Variation (0.01449201) is less the Alpha (0.05) so we can say that the closing method employed did make a difference in the sales with no more than a 5% chance of being wrong. The alpha can be thought of as the maximum chance of being wrong.

Hand calculations of this entire problem are shown at the end of this module. After taking a quick look at that, you will probably be happy to let Excel do all of the calculation work. The calculation of the P value and the associated F Value are time-consuming and rigorous. No need to do them by hand if Excel will do it for you.

ANOVA: Two Factor Without Replication

Problem 2: Three Sales Closing Methods, Five Salespeople, and Two-Factor ANOVA Without Replication

Problem: A sales manager was experimenting to determine the effectiveness of 3 different types of closing methods. He instructed each of his 5 salespeople to try all 3 methods. Each salesperson would use only one of the closing methods for one full week. Every week each salesperson had to switch to another of the 3 closing methods that he had not previously used. At the end of 3 weeks, the sales results were provided below. Determine with 99% certainty whether the closing methods used made a difference in the sales results. Also determine with 99% certainty whether the individual abilities of each salesperson made a difference in the sales results.

	Method 1	Method 2	Method 3
Salesperson 1	51	57	72
Salesperson 2	109	112	117
Salesperson 3	47	43	51
Salesperson 4	98	98	107
Salesperson 5	70	69	77

Note that this is different than the test run for Single factor ANOVA above. The single factor ANOVA test used 12 salespeople who each used only one closing method. This two-factor ANOVA uses only five salespeople. Each salesperson here must use each of the three closing methods for one week.

Problem Solving Steps

This is a **Two Factor ANOVA** test because there are two variables that could effect the measured outcome (sales). The two given variables are the selling ability of each salesperson and the closing method used. Nothing else is entered into this test that might have an effect on the measured output (sales).

This is a test **Without Replication** because there is no way to test whether the two variables interact with each other to affect the measured outcome. An example of a test With Replication is in the ANOVA example that follows this one.

The two **Null Hypotheses** for this test are as follows. Each of the two variables (the salesperson's ability and the closing method used) have their own Null Hypothesis. The Null Hypothesis for salesperson's ability states that there is not enough variation in the abilities of the salespeople to influence sales. In other words, the measured output (sales) will not be affected by which salesperson does the selling. The Null Hypothesis for the closing method used states that the choice of closing method will have no affect on sales.

Excel Instructions:

Tools / Data Analysis / Anova: Two-Factor Without Replication (in Excel 2010 -> Data tab)

This brings up the following dialog box:

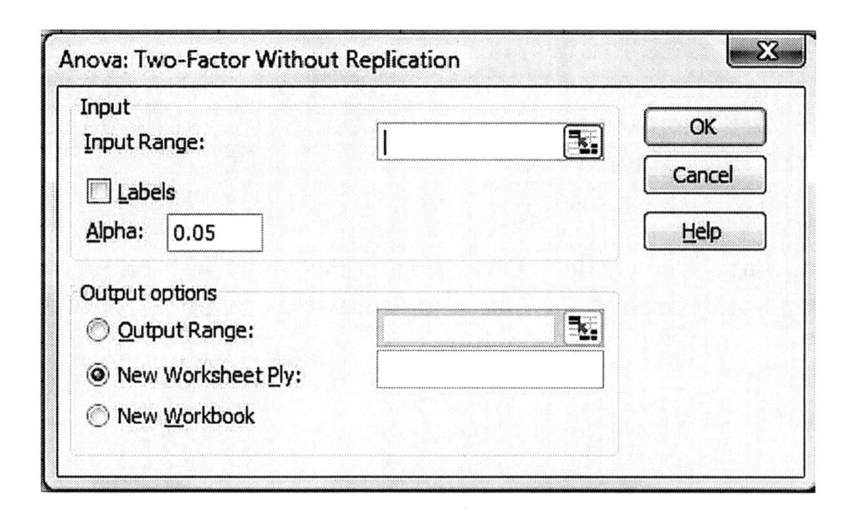

You will be required to provide the following 5 inputs:

1) Input Range: Highlight the data and labels as is highlighted in Yellow previously. Note that your data must be initially arranged in columns as above or similarly in rows. Note that both the Method labels and the Salesperson labels are highlighted because they are the two factors being analyzed.

2) Grouped By: Select Columns or Rows, depending on how your data is arranged. In this case the data is arranged in columns

3) Data Labeling: Select whether your data is labeled in the first column (if the data is grouped in rows) or labeled in the first row (if your data is grouped in columns). In this case, the data is labeled in the first row.

4) Alpha: You will type in the level of significance here. The level of significance (alpha) is the max level of error you are willing to tolerate. If you can only tolerate a possibility of 1% of error, the level of significance (alpha) here would be 0.01. Prefilled in this input box is 0.05. This represents a maximum error of 5%. Another way of looking at this would be to say that 5% maximum error represents being at least 95% certain of the correctness of the output. Type in 0.01 for Alpha.

5) Output: Here you would enter (or select) the upper-left corner of where you want the output printed on your worksheet. You will need to make sure that the place that you have specified the output to be printed has at least 7 free columns and 20 free rows.

Hit "OK" and Excel will perform the Two-Factor ANOVA Without Replication analysis and output the following result is shown in the location that you have specified.

On the next page is the output:

Anova: Two-Factor Without Replication

SUMMARY	Count	Sum	Average	Variance
Salesperson 1	3	180	60	117
Salesperson 2	3	338	112.6666667	16.33333333
Salesperson 3	3	141	47	16
Salesperson 4	3	303	101	27
Salesperson 5	3	216	72	19
Method 1	5	375	75	767.5
Method 2	5	379	75.8	819.7
Method 3	5	424	84.8	724.2

ANOVA

Source of Variation	SS	df	MS	F	P-value	F crit
Rows	9151.066667	4	2287.766667	193.6050776	5.42004E-08	7.006076623
Columns	296.1333333	2	148.0666667	12.5303244	0.003428581	8.649110641
Error	94.53333333	8	11.81666667			
Total	9541.733333	14				

Alpha = 0.01

The p-Value for the Rows (5.42004E-08) is much less than the level of significance (0.05) so the individual selling ability of each salesperson made a difference in sales results.

The p-Value for columns (0.003428581) is much less than the level of significance (0.05) so the closing method used made a difference in sales results

Analyze the Output

The P-value

Without going into too much statistical detail, the cells in the output entitled P-value will tell you whether the three different closing methods and individual selling abilities have an effect on the measured output of sales

General Rule:

If P-value < alpha ----> The different treatments affected the output

If P-value > alpha ----> The different treatments did not affect the output

The data interpretation for Two-Factor ANOVA is very similar to Single Factor ANOVA. The only difference is that now two variables (choice of salesperson and choice of closing method) are being tested to determine whether either had an effect on the measured output (sales). If the P-value associated with a variable is less than alpha (0.01), then we can state that this variable had an effect on sales.

Data for each salesperson is arranged in rows. The P-value associated with choice of salesperson is highlighted darker. This is the P-value for the source of variation in rows. This P-value is 5.42004 x 10-8 and is much less than Alpha (0.01) so we can state that the choice of salesperson almost certainly had an effect on sales.

Data for each closing method is arranged in columns. The P-value associated with choice of closing method is highlighted lighter. This Is the P-value for the source of variation in columns. This P-value is .003428581 and is much less than alpha (0.01) so we can state that the choice of closing method almost certainly had an effect on sales.

Anova: Two Factor With Replication

Problem 3: Three Headlines, Three Ad Texts, their Interaction, and Two-Factor ANOVA With Replication

Problem: A company was conducting pay-per-click advertising on both Google and Yahoo. The Internet marketing manager was testing headlines and ad text for ads he had written for one product. The manager created 3 headlines and 3 sets of ad text. Altogether there were 9 possible ad combinations of headline and ad text. The manager then ran all 9 ads for an equal number of times on both the Google and Yahoo paid search networks. Below is the resulting Click-Through Rate (CTR) for each combination of Headline / Ad Text / Search Engine after the ads ran for 2 weeks. Use a 95% confidence level to determine whether 1) Ad Text, 2) Headline, and 3) Interaction between Ad Text and Headline had an effect on Click-Through Rate.

	Headline 1	Headline 2	Headline 3	
Ad Text 1	2.80	2.04	1.58	Google
	2.73	1.33	1.26	Yahoo
Ad Text 2	3.29	1.50	1.00	Google
	2.68	1.40	1.82	Yahoo
Ad Text 3	2.54	3.15	1.92	Google
	2.59	2.88	1.33	Yahoo

Problem Solving Steps

This is a **Two Factor ANOVA** test because there are two variables that could effect the measured outcome (Click-Through Rate on each search engine) The two given variables are the Ad Text and Headline.

This is a test **With Replication** because the same tests are performed in different places to determine if there is interaction between the variables (Ad Text and Headline) that might affect the measured output (Click-Through Rate on each search engine).

Null Hypotheses for all variables - Each of the two variables (Ad Text and Headline) has its own Null Hypothesis. The Null Hypothesis for Ad Text states that there is no variation in Click-Through Rate on each search engine as a result of different Ad Text. In other words, the measured output (CTR in each search engine) will not be affected by the Ad Text used in the ads. The Null Hypothesis for Headline states that the choice of Headline for ads in each search engine will not affect Click-Through Rate for that search engine.

The replication of the tests in different search engines provides the chance to test whether interaction between the two variables affects the measured outcome. The Null Hypothesis associated with this facet of the test states that interaction between the two variables will not have an affect on the output.

Excel Instructions:

Tools / Data Analysis / Anova: Two-Factor With Replication (in Excel 2010 -> Data tab)

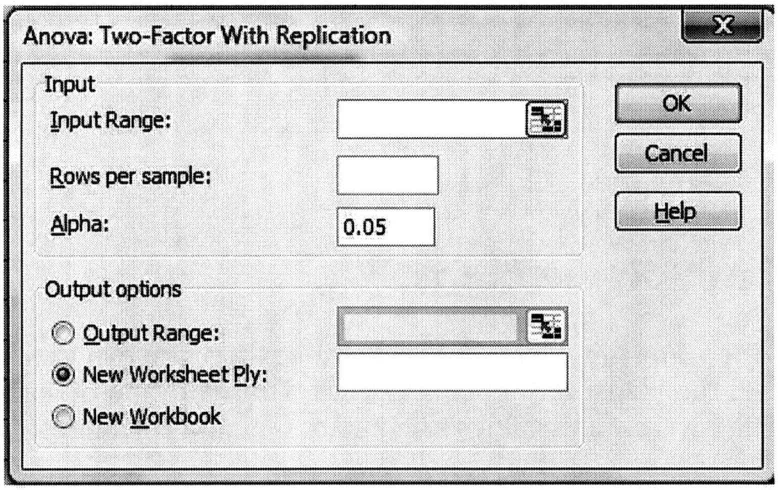

Anova: Two-Factor With Replication
Input
Input Range:
Rows per sample:
Alpha: 0.05
Output options
Output Range:
New Worksheet Ply:
New Workbook
OK
Cancel
Help

You will be required to provide the following 4 inputs:

1) Input Range: Highlight the data and labels as is highlighted in Yellow previously. Note that your data must be initially arranged in columns as above or similarly in rows. Note that both the Headline labels and the Ad Text labels are highlighted because they are the two factors being analyzed. Also note that data from duplicate tests run in different territories is arranged in pairs of rows, not columns. Excel requires this.

You do not need to state whether your columns and rows are labeled. One caution - if you use labels, place labels on both the columns and rows, as is the case in this example. Both columns and rows need to be labeled, or do not use labels at all.

2) Rows per Sample - State how many times each test will be duplicated in different locations. In this case, Each test was duplicated in one additional territory, so there are 2 Rows per Sample in this example.

3) Alpha: You will type in the level of significance here. The level of significance (Alpha) is the max level of error you are willing to tolerate. If you can only tolerate a possibility of 1% of error, the level of significance (alpha) here would be 0.01. Prefilled in this input box is 0.05. This represents a maximum error of 5%. Another way of looking at this would be to say that 5% maximum error represents being at least 95% certain in the correctness of the output.

4) Output: Here you would enter (or select) the upper-left corner of where you want the output printed on your worksheet. You will need to make sure that the place where you have specified the output to be printed has at least 7 free columns and 36 free rows.

Hit "OK" and Excel will perform the Two Factor ANOVA with Replication analysis and output the following result on the next page.

Anova: Two-Factor With Replication

SUMMARY	Headline 1	Headline 2	Headline 3	Total
Ad Text 1				
Count	2	2	2	6
Sum	5.53	3.37	2.84	11.74
Average	2.765	1.685	1.42	1.956666667
Variance	0.00245	0.25205	0.0512	0.467226667
Ad Text 2				
Count	2	2	2	6
Sum	5.97	2.9	2.82	11.69
Average	2.985	1.45	1.41	1.948333333
Variance	0.18605	0.005	0.3362	0.750576667
Ad Text 3				
Count	2	2	2	6
Sum	5.13	6.03	3.25	14.41
Average	2.565	3.015	1.625	2.401666667
Variance	0.00125	0.03645	0.17405	0.444776667
Total				
Count	6	6	6	
Sum	16.63	12.3	8.91	
Average	2.771666667	2.05	1.485	
Variance	0.073256667	0.62848	0.12407	

ANOVA

Source of Variation	*SS*	*df*	*MS*	*F*	*P-value*	*F crit*
Sample	0.807211111	2	0.403605556	3.477026898	0.076062669	4.256494729
Columns	4.991077778	2	2.495538889	21.49885134	0.00037339	4.256494729
Interaction	2.277122222	4	0.569280556	4.904302671	0.022409686	3.633088512
Within	1.0447	9	0.116077778			
Total	9.120111111	17				

Alpha = 0.05

The p-Value for Sample (0.076062) is more than the level of significance (0.05). We cannot reject the NULL Hypothesis that states that the Ad Text does not affect Click-Through Rate.

The p-value for Columns (0.00037339) is less than the level of significance (0.05). This indicates that choice of Headline affects Click-Through Rate. This is a rejection of the Null Hypothesis

The p-Value for Interaction (0.022409) is less than the level of significance. This indicates that different combinations of interactions (Ad Text / Headline) affects CTR.

Analyze the Output

The P-value

Interpreting the Output: Without going into too much statistical detail, the cells in the output entitled P-value will tell you whether the different ad text, headlines, and interactions between them had an effect on the measured output (Click-Through Rate on the two search engines). Highlighted as light gray (lightest highlighting and highest position), medium gray (medium highlighting and middle position), and dark gray (darkest highlighting and bottom position) on the Excel output is the P-value for each of these variables.

General Rule:

If P-value < alpha ----> The different treatments affected the output

If P-value > alpha ----> The different treatments did not affect the output

The data interpretation for Two-Factor ANOVA is very similar to Single Factor ANOVA. The only difference is that now two variables (choice of ad text and choice of headline) along with interaction between the two variables are being tested to determine whether any of these has an effect on the measured output (Click-Through Rate). If the P-value associated with any of these variable is less than alpha (0.05), then we can state that this variable had an effect on the measured output (CTR).

Data for each Ad Text is arranged in rows. Excel has labeled this "Samples" in the output. The P-value associated with Ad Text is highlighted in the lightest gray and top position. This is the P-value for the source of variation in rows. This P-value is 0.076062 and is GREATER than alpha (0.05) so we can state with no more than 5% chance of error that Ad Text did not affect measured output (Click-Through Rate in each search engine). In other words, we cannot disprove the Null Hypothesis associated with this variable.

Data for each Headline is arranged in columns. The P-value associated with choice of Headline is highlighted medium gray and in middle position. This is the P-value for the source of variation in columns. This P-value is .00037339 and is much less than Alpha (0.05) so we can state that the choice of Headline almost certainly had an effect on the measured output (Click-Through Rate in each search engine).

Data and the P-Value associated with possible interaction between the two variables is highlighted in dark gray and on bottom position. This P-value is 0.022409688 and is less then the alpha (0.05) so we can state with less than 5% chance of error that interaction between the variables (Ad Text and Headline) did have an effect on Click-Through Rate

Hand Calculation of ANOVA: Single Factor

(Excel calculation of Single Factor ANOVA is shown at the top of this Worksheet)

Problem 4: Hand Calculation of Closing Methods and Single Factor ANOVA

Problem: Three different sale closing methods were used. Three groups of four salespeople were randomly chosen. Each group was instructed to use only one of the closing methods for all of their sales. Sales totals of each salesperson over the next two weeks were collected. Determine with a 95% level of certainty whether there is a difference in the effectiveness of the closing methods. Below are the sales results for all 12 salespeople. All salespeople were assumed to have equal closing abilities.

The data is shown on the following page:

Sales Group 1	Closing Method 1
Salesperson 1	16
Salesperson 2	21
Salesperson 3	18
Salesperson 4	13

Sales Group 2	Closing Method 2
Salesperson 5	19
Salesperson 6	20
Salesperson 7	21
Salesperson 8	20

Sales Group 3	Closing Method 3
Salesperson 9	24
Salesperson 10	21
Salesperson 11	22
Salesperson 12	25

Arrange the data as below to facilitate calculations:

Sales Group ---->	Group 1	Group 2	Group 3
Treatment ---->	Method 1	Method 2	Method 3
Salesperson A	16	19	24
Salesperson B	21	20	21
Salesperson C	18	21	22
Salesperson D	13	20	25

Here over the next several pages are the Hand-Calculations for this Single Factor ANOVA problem:

Column Total	68	80	92
Column Mean	17	20	23
Grand Mean = (17 + 20 + 23) / 3			
Grand Mean =	20		
Column Mean - Grand Mean	-3	0	3
(Column Mean - Grand Mean)^2	9	0	9
# Rows * [(Column Mean - Grand Mean)^2]	36	0	36
Sum of Squares Between Groups = 36 + 0 + 36 = 72			

Method 1	Method 2	Method 3
16	19	24
21	20	21
18	21	22
13	20	25
68	80	92
17	20	23

Method 1	Method 2	Method 3
16 - 17	19 - 20	24 - 23
21-17	20 - 20	21 - 23
18 - 17	21 - 20	22 - 23
13 - 17	20 - 20	25 - 23

Method 1	Method 2	Method 3
-1	-1	1
4	0	-2
1	1	-1
-4	0	2

Square each

Method 1	Method 2	Method 3
1	1	1
16	0	4
1	1	1
16	0	4
34	2	10

Sum of Squares Within Treatments = 34 + 2 + 10 = 46

Degrees of Freedom

Between Groups DOF = # groups - 1 = c - 1 = 3 - 1 = 2

Within Groups DOF = C(r-1) = 3 (4 - 1) = 9

Total Degrees of Freedom = 9 + 2 = 11

Sum of Squares

Between Groups Sum of the Squares	72
Sum of Squares Within Groups	46
Total Sum of the Squares	118

Mean Squares

MS = Mean Square = Sum of Squares / degrees of freedom

SS	*df*	*MS*
72	2	36
46	9	5.111111111

F Statistic

F Statistic = (MS Between Group) / (MS Within Groups)

F Statistic = 36 / 5.111111 = 7.043478261

p Value

p-Value = FDIST(F Statistic,DOF Between Groups,DOF Within Groups) =

p-Value = FDIST(7.043478,2,9) =
0.0144192029269803

> The p-value of 0.014419 is less than the designated level of significance of 0.05. This indicates that there is less than a 5% chance that this result could have occurred if there was no difference in effectiveness between the methods. Therefore, there is at least 95% certainty that there is a real difference in effectiveness of the methods. The Null Hypothesis which was rejected states that choice of closing method does not affect sales.

The p-Value represents the proportion of area under the F Distribution curve to the right of the given F value. If this p-Value is less than the stated level of significance, this demonstrates that there is a difference in the objects or process being analyzed. - in other words, there is a difference in the variances.

Chapter 13 - Chi-Square Independence Test

Basic Explanation of the Chi-Square Independence Test

The Chi-Square Independence Test is used to determine whether two attributes of one object are independent of each other. For example, The Chi-Square Independence Test could be used to determine whether salary and years of education are independent of each other for a sampled group of people.

A large random sample is taken from a population. Each object sampled is measured for two attributes. Each sampled object is then placed in a like group of other similar sampled objects that display the same measurements for each of the two attributes. A Chi-Square statistic, X^2, is calculated from all of the groups combined.

The calculated Chi-Square statistic is compared with the critical Chi-Square statistic. The critical Chi-Square statistic is an independent number determined by the problem's degrees of freedom and Level of Significance.

If the calculated Chi-Square statistic is greater than the critical Chi-Square statistic, then the two attributes are not independent of each other.

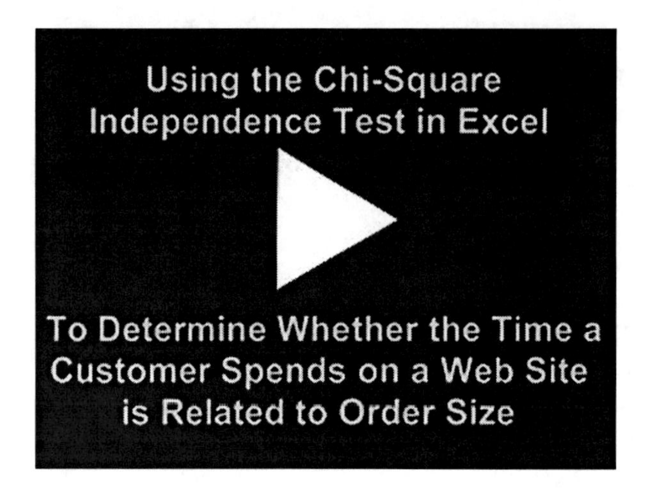

Instructional Video
Go to
http://www.youtube.com/watch?v=5wYikibiVmg
to View a
Video From Excel Master Series
About How To Perform the
Chi-Square Independence
Test in Excel

(Is Your Internet Connection and Sound Turned On?)

Level of Certainty

The Level of Certainty = 1 - **α** = Degree of certainty required by the problem. If the problem requires 95% certainty, then the Level of Significance, **α**, equals 0.05.

Level of Significance

The Level of Significance, **α**, is equivalent to the maximum possibility of error. The Level of Significance can also be derived from the Required Level of Certainty as shown above.

Contingency Table

The Contingency Table contains the data showing how the two attributes are distributed among the sampled objects. One attribute has its different classifications grouped in rows. The other attribute has its different classifications grouped in columns. If, for example, one attribute has three levels and the other attribute has three levels, then the Contingency Table has nine cells. Each cell corresponds to one of the nine possible combinations of the three levels of one attribute and the three levels of the other attribute. Each data sample will be placed in one of the nine cells, depending on which combination of the three levels of the two attributes that it possesses. The Contingency Table described here is a 3x3 table. It has 3 rows and 3 columns.

Degrees of Freedom

Degrees of Freedom, **v**, for a contingency table with **r** number of rows and c number of columns equals the following:

v = (**r** - 1) * (**c** - 1)

For example, if one attribute has three levels and the other attribute also has three attributes, the Contingency Table will have three rows and three columns. In this case, the Degrees of Freedom equals:

v = (**r** - 1) * (**c** - 1) = (3 - 1) * (3 - 1) = 2 * 2 = 4

Chi-Square Distribution

The Chi-Square distribution is actually a number of different distributions. The Chi-Square distribution has only one parameter, its degrees of freedom, abbreviated as either **v** or df. Different degrees of freedom produce a different-shaped Chi-Square distribution. Overall, the Chi- Square is somewhat bell-shaped with the "bell" skewed to the left. The Chi-Square distribution begins at 0 as its lowest value. It rapidly rises to the peak of its "bell" shape and then gradually tapers to an outer right tail as distribution values get larger.

As the degrees of freedom increase, the peak of the "bell" curve moves to the right and flattens out. Each different number of degrees of freedom produces a unique`+ Chi-Square distribution.

Critical Chi-Square Statistic

The Critical Chi-Square Statistic can be calculated given the degrees of freedom and the Level of Significance. These two parameters are obtained from the problem. The degrees of freedom are determined by how many sampling possibilities exist. For example, a 3x3 Contingency Table would generate 4 degrees of freedom [**v** = (**r** - 1) * (**c** - 1)]. Four degrees of freedom corresponds to its own unique Chi-Square distribution. The Level of Significance is specified by the problem. If the problem requires a Level of Certainty of 95%, then the Level of Significance, **α**, is 0.05.

The Level of Significance, **α**, is closely related to the Critical Chi-Square Statistic in the following way:

The Level of Significance, α, represents the area under the Chi-Square distribution curve that is to the right of the Critical Chi-Square statistic.

If a problem's calculated Chi-Square statistic is less than the Critical Chi-Square statistic for the problem's given Level of Significance and degrees of freedom, then the object's tested attributes are independent (at least to the degree of certainty required by the problem).

If a problem's calculated Chi-Square statistic is greater than the Critical Chi-Square statistic, then the object's tested attributes are not independent.

Independence Test Rule:

If Calculated Chi-Square Value > Critical Chi-Square Statistic ----> Attributes NOT Independent

If Calculated Chi-Square Value < Critical Chi-Square Statistic ----> Attributes ARE independent

Or Equivalently:

If α > p Value ----> Attributes NOT Independent

If α < p Value ----> Attributes ARE Independent

Excel Functions Used When Performing the Chi-Square Independence Test

CHIINV (Level of Significance, Degrees of Freedom)

= $X^2_{(\alpha, v)}$ = **Critical Chi-Square Statistic**

For example, if a problem requires a 95% Confidence Level, and the Contingency Table has 3 rows and 3 columns, then the Level of Significance, **α** = 0.05, and the Degrees of Freedom = 4

[**v** = (**r** - 1) * (**c** - 1) = (3-1) * (3-1) = 2 * 2 = 4]

Critical Chi-Square Statistic = $X^2_{(\alpha=0.05,\ v=4)}$ = CHIINV (0.05,4) = 9.488

CHIDIST (<u>Critical</u> Chi-Square Statistic, Degrees of Freedom)

= **Level of Significance (α)**

= Area under the Chi-Square Distribution Curve to the Right of the Critical Statistic

For example:

CHIDIST (9.488, 4) = 0.05

-----> **α** = 0.05

-----> The area under the Chi-Square Distribution Curve to the right of 9.488 equals 0.05 (5%) of the total area under the curve. The outer right tail to the right of 9.488 contains 5% of the total area under the curve.

CHIDIST (<u>Calculated</u> Chi-Square Statistic, Degrees of Freedom)

= p Value

= Area under the Chi-Square Distribution Curve to the Right of the Calculated Chi-Square Statistic

Problem: Determine if There is a Relationship Between the Time in a Store and the Number of Items Purchased

Problem: A large department store wanted to determine if there was any relationship between the time that a customer would spend in the store and the number of items that the customer bought. A random sample of 10,000 customers was taken. Determine with a maximum error of 1% whether those two customer actions are independent.

Data and the resulting Contingency Table for this problem are shown on the next page.

Contingency Table with Data from 10,000 Randomly Sampled Customers

Length of Time in Store		Number of Items Purchased			
		Zero	One	Two	Total
	0 to 10 minutes.	1000	900	100	2000
	10 to 20 minutes	1500	2600	500	4600
	More than 20 minutes	500	2500	400	3400
	Total	3000	6000	1000	10000

Given these totals for each row and column:

Length of Time in Store		Number of Items Purchased			
		Zero	One	Two	Total
	0 to 10 minutes.				2000
	10 to 20 minutes				4600
	More than 20 minutes				3400
	Total	3000	6000	1000	10000

We would expect to see this distribution of samples data in the cells if the attributes of number of items purchased and length of time in store were independent:

Length of Time in Store		Number of Items Purchased			
		Zero	One	Two	Total
	0 to 10 minutes.	600	1200	200	2000
	10 to 20 minutes	1380	2760	460	4600
	More than 20 minutes	1020	2040	340	3400
	Total	3000	6000	1000	10000

These numbers were calculated as follows:

Length of Time in Store		Number of Items Purchased			
		Zero	One	Two	Total
	0 to 10 minutes.	600 = 3000*(2000 / 10000)	1200 = 6000*(2000 / 10000)	200 = 1000*(2000 / 10000)	2000
	10 to 20 minutes	1380 = 3000*(4600 / 10000)	2760 = 6000*(4600 / 10000)	460 = 1000*(4600 / 10000)	4600
	More than 20 minutes	1020 = 3000*(3400 / 10000)	2040 = 6000*(3400 / 10000)	340 = 1000*(3400 / 10000)	3400
	Total	3000	6000	1000	10000

Calculation of the Chi-Squared Statistic for the Sample Data

	Observed Number of Customers	Expected Number of Customers			
	f_o	f_t	$f_o - f_t$	$(f_o - f_t)^2$	$(f_o - f_t)^2 / f_t$
	1,000	600	400	160,000	266.7
	1,500	1,380	120	14,400	10.4
	500	1,020	-520	270,400	265.1
	900	1,200	-300	90,000	75
	2,600	2,760	-160	25,600	9.3
	2,500	2,040	460	211,600	103.7
	100	200	-100	10,000	50
	500	460	40	1,600	3.5
	400	340	60	3,600	10.6
Total	10,000	10,000	0		794.3
					X^2 = 794.3

The Chi-Square statistic for this sample = X^2 = 794.3

Level of significance (maximum error) = α = 0.01

Number of Rows in the Contingency Table = **r** = 3

Number of Columns in the Contingency Table = **c** = 3

Degrees of Freedom = **v** = (**r** - 1) * (**c** - 1) = (3 - 1) * (3 - 1) = 4

Critical Chi-Square Statistic = $X^2_{(\alpha=0.01, v=4)}$

= **CHINV** (0.01, 4) = **13.28**

Chi-Square Statistic for this Sample = X^2 = 794.3

Critical Chi-Square Statistic = $X^2_{(\alpha=0.01,\ v=4)}$ = 13.28

p Value = CHIDIST (Calculated Chi-Square Statistic, v)

= CHIDIST(794.3,4) ≈ 0

α = 0.01

Independence Test Rule:

If Calculated Chi-Square Value>Critical Chi-Square Statistic ----> Attributes NOT Independent

If Calculated Chi-Square Value<Critical Chi-Square Statistic ----> Attributes ARE independent

Or Equivalently:

If α>p Value ----> Attributes NOT Independent

If α<p Value ----> Attributes ARE Independent

Since the Chi-Square statistic from the sample data (X^2= 794.3) exceeds the Critical value of Chi-Square = $X^2(\alpha=0.01, v=4)$= 13.28, we can conclude that the two customer actions (length of store stay and number of items purchased) are not independent. This result can be more clearly understood be examining the following diagram on the next page.

The following diagram illustrates a Chi-Square distribution curve with 4 degrees of freedom. The outer 1% of the total area under this curve begins at the Critical Chi-Square Statistic, which is 13.28 as shown in the diagram below.

Critical Chi-Square Statistic = $X^2_{(\alpha=0.01,\ v=4)}$

= **CHINV** (0.01, 4) = **13.28**

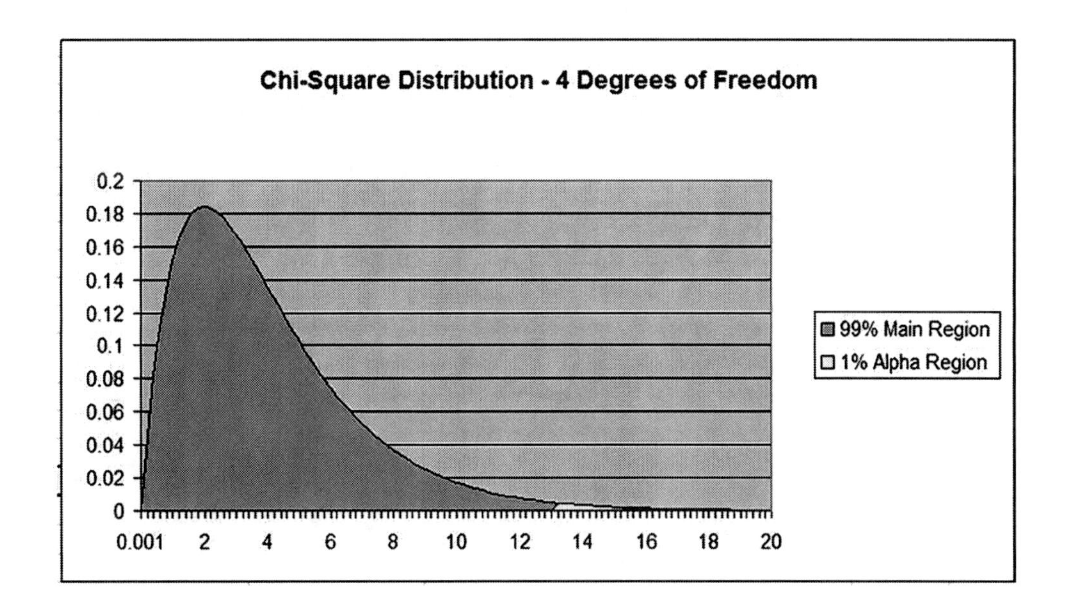

Chapter 14 - Chi-Square Population Variance Test

Basic Explanation of the Chi-Square Population Variance Test

This Chi-Square test tells whether the variance of a population has changed. This test is quite often used to check processes to determine if the process' variance has changed. A common usage of this test is to check whether a production line process has had any change in variance.

The Chi-Square Distribution is used to determine if a population's variance has changed. Quality control people use the Chi-Square test to determine if a process' variance levels are staying within given limits.

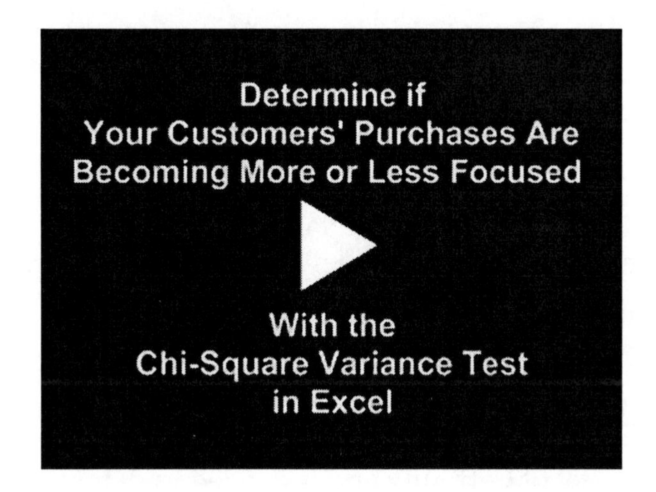

Instructional Video

Go to
http://www.youtube.com/watch?v=IrVAqAvsxBQ
to View a
Video From Excel Master Series
About How To Perform the
Chi-Square Population Variance
Test in Excel

(Is Your Internet Connection and Sound Turned On?)

The 5-Step Chi-Square Population Variance Test

For a process that has a known standard deviation, σ, perform these five steps to determine if the standard deviation, and therefore the variance, has changed ----> Variance = (Standard Deviation)^2

1st Variance Test Step - Determine the Level of Certainty and α

Determine the Level of Certainty and **α** needed to determine whether the variance has changed.

----> Level of Certainty = 1 - **α**

2nd Variance Test Step - Measure Sample Standard Deviation, s, from a large sample (n>30)

Measure Sample Standard Deviation, s, from a large sample (**n**>30)

3rd Variance Test Step - Calculate the Chi-Square Statistic

Chi-Square Statistic, = **[** (**n**-1)*(**s***s) **]** / (**σ***σ)

4th Variance Test Step - Calculate the Curve Area Outside of the Chi-Square Statistic

a) If Sample Standard Deviation, **s**, **is greater than** the population Standard Deviation, **σ**, then:

Calculate the Area in the Right Outer Tail to the Right of the Chi-Square Statistic

Tail Area Right of Chi-Square Statistic = CHIDIST(Chi-Square Statistic, **n**-1)

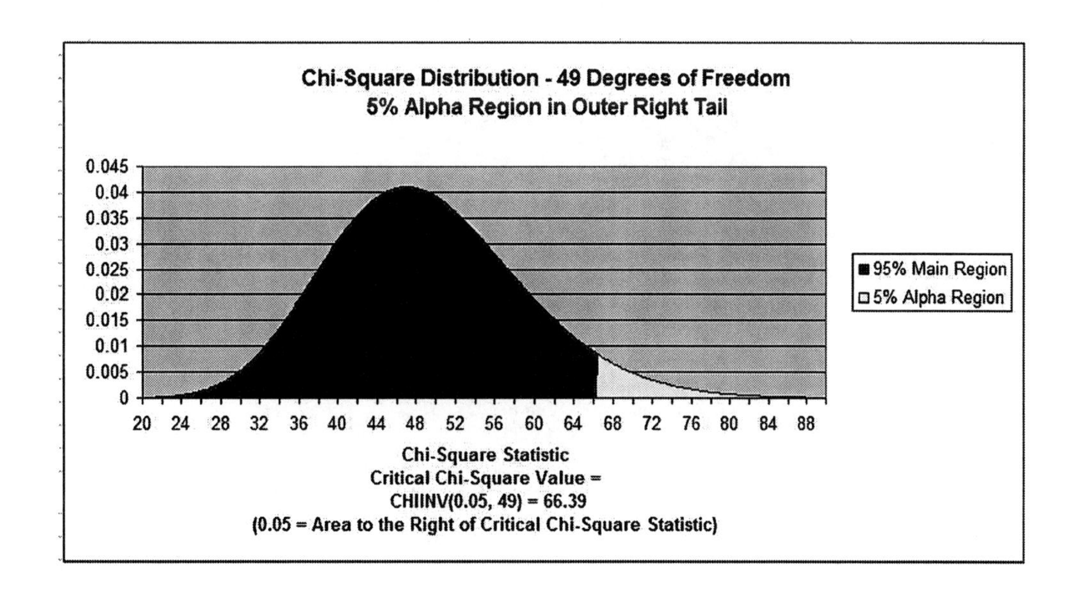

b) If Sample Standard Deviation, **s**, **is less than** the population Standard Deviation, σ, then:

Calculate the Area in the Left Outer Tail to the Left of the Chi-Square Statistic

Tail Area Left of Chi-Square Statistic = 1 - CHIDIST(Chi Square Statistic, **n**-1)

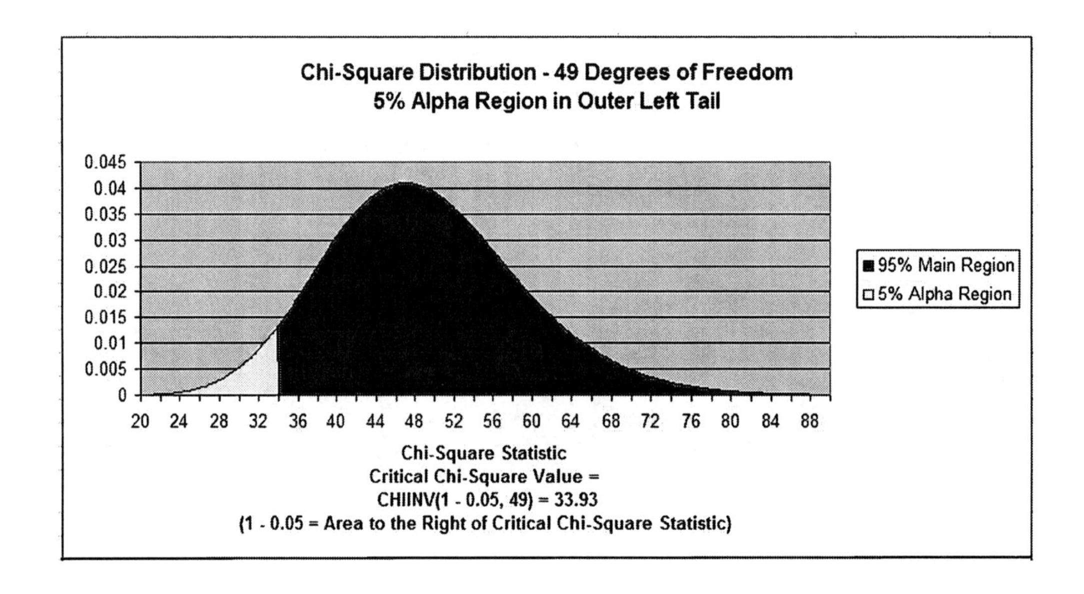

5th Variance Test Step - Analyze Using the Chi-Square Statistic Rule

Rule: If the Curve Area Outside of the Chi-Square Statistic is less than the Level of Significance, α, the Variance has moved in the direction toward Sample Standard Deviation.

Problem: Use the Chi-Square Population Variance Test to Determine if a Population Variance Has Increased

Problem: A manufacturer wants to check if the variance on a process has changed. A machine drills a hole as part of the manufacturing process. The standard deviation of the hole diameter has historically been 1.6 ml. A random sample of 50 hole diameters were checked in one batch. The measured sample standard deviation was 1.9 ml. At a 0.05 level of significance, has the population standard deviation increased above 1.6 ml?

<u>Apply the 5-Step Chi-Square Variance Change Test</u>

1st Variance Test Step - Determine the Level of Certainty and α needed to determine whether the variance has changed.

----> Level of Certainty = 1 - **α** = 1 - 0.05 = .95 = 95%

α = 0.05

2nd Variance Test Step - Measure Sample Standard Deviation, **s**, from a large sample (**n**>30)

s = 1.9

3rd Variance Test Step - Calculate the Chi-Square Statistic

Chi-Square Statistic, = **[** (**n**-1)*(**s*****s**) **]** / (**σ*****σ**)

Chi-Square Statistic = **[** $(50-1)*(1.9)^2$ **]** / $(1.6)^2$

= 69.09766

4th Variance Test Step - Calculate the Curve Area Outside of the Chi-Square Statistic

a) If Sample Standard Deviation, s, is **Greater Than** the Population Standard Deviation, **σ**, **(the case here)** then:

Tail Area Right of Chi-Square Statistic

= CHIDIST(Chi-Square Statistic, **n**-1)

= CHIDIST(69.09766, 49) = **0.030749**

b) If Sample Standard Deviation, s, **is less than** the population Standard Deviation, **σ**, **(not the case here)** then:

Tail Area Left of Chi-Square Statistic

= **1 -** CHIDIST(Chi-Square Statistic, **n**-1)

5th Variance Test Step - Analyze Using the Chi-Square Statistic Rule

Rule: If the Curve Area Outside of the Chi Square Statistic is less than the Level of Significance, α, the Variance has moved in the direction toward Sample Standard Deviation.

Tail Area Right of Chi-Square Statistic

= CHIDIST(69.09766, 49) = 0.030749

Level of Significance = **α** = 0.05

In this case, the Tail Area to the Right of the Chi-Square Statistic (0.030749) is less than the Level of Significance (0.05) so we can state with 95% certainty that the population standard deviation has increased.

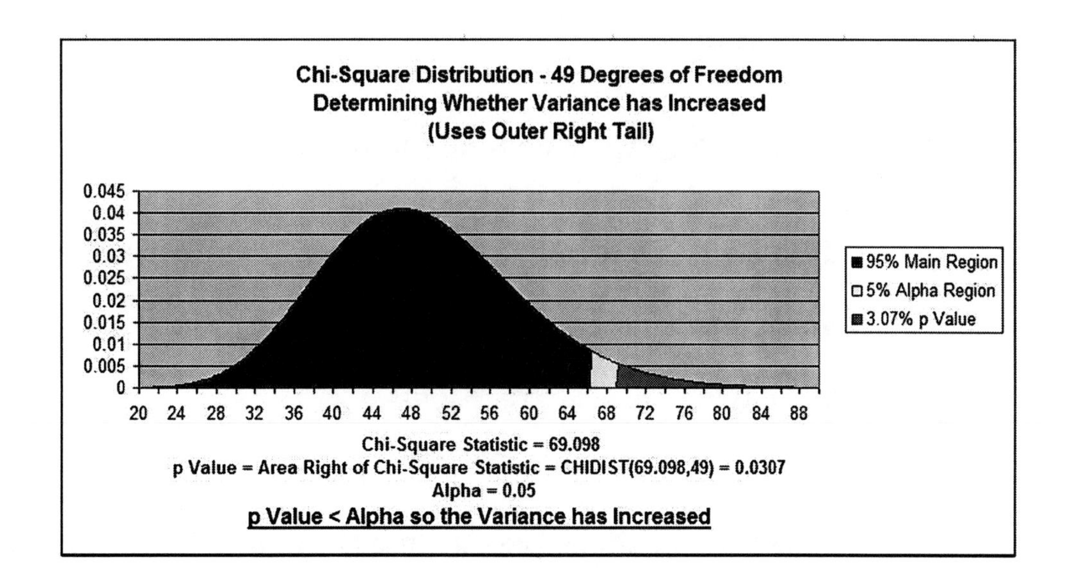

Problem: Use the Chi-Square Population Variance Test to Determine if a Population Variance Has Decreased

Problem: A manufacturer wants to check whether the variance on a process has changed. A machine drills a hole as part of the manufacturing process. The standard deviation of the hole diameter has historically been 1.6 ml. A random sample of 50 hole diameters were checked in one batch. The measured sample standard deviation was 1.375 ml. At a 0.05 level of significance, has the population standard deviation decreased below 1.6 ml?

Sample Size = n = 50

Degrees of Freedom = n - 1 = 49

Level of Significance = α = 0.05

Population Standard Deviation = σ = 1.6

Sample Standard Deviation = s = 1.375

Apply the 5-Step Chi-Square Variance Change Test

1st Variance Test Step - Determine the Level of Certainty and α

needed to determine whether the variance has changed.

----> Level of Certainty = 1 - **α** = 1 - 0.05 = .95 = 95%

α = 0.05

2nd Variance Test Step - Measure Sample Standard Deviation, **s**, from a large sample (**n**>30)

s = 1.375

3rd Variance Test Step - Calculate the Chi-Square Statistic

Chi-Square Statistic, = **[** (**n**-1)*(**s*s**) **]** / (**σ*σ**)

Chi-Square Statistic = **[** (50-1)*(1.375)² **]** / (1.6)²

= 36.18774

4th Variance Test Step - Calculate the Curve Area to the Outside of the Chi-Square Statistic

a) If Sample Standard Deviation, s, **is greater** than the population Standard Deviation, σ, **(not the case here)** then:

Tail Area Right of Chi Statistic = CHIDIST(Chi Square Statistic, **n**-1)

b) If Sample Standard Deviation, s, **is less** than the population Standard Deviation, σ, **(the case here)** then:

Tail Area Left of Chi Statistic =

= **1** - CHIDIST(Chi-Square Statistic, n-1)

= **1** - CHIDIST(36.18774, 49)

= **1** - 0.912951

= 0.087049

5th Variance Test Step - Analyze Using the Chi-Square Statistic Rule

Rule: If the Curve Area to the Outside of the Chi-Square Statistic is less than the Level of Significance, α, the Variance has moved in the direction toward Sample Standard Deviation.

Left Tail **Area Outside of Chi Statistic**

= 1 - CHIDIST(36.18774, 49) = 0.087049

Level of Significance = **α** = 0.05

In this case, the Left Area to the Outside of the Chi-Square Statistic (0.087049) is greater than the Level of Significance (0.05) so we cannot state that the population standard deviation has decreased.

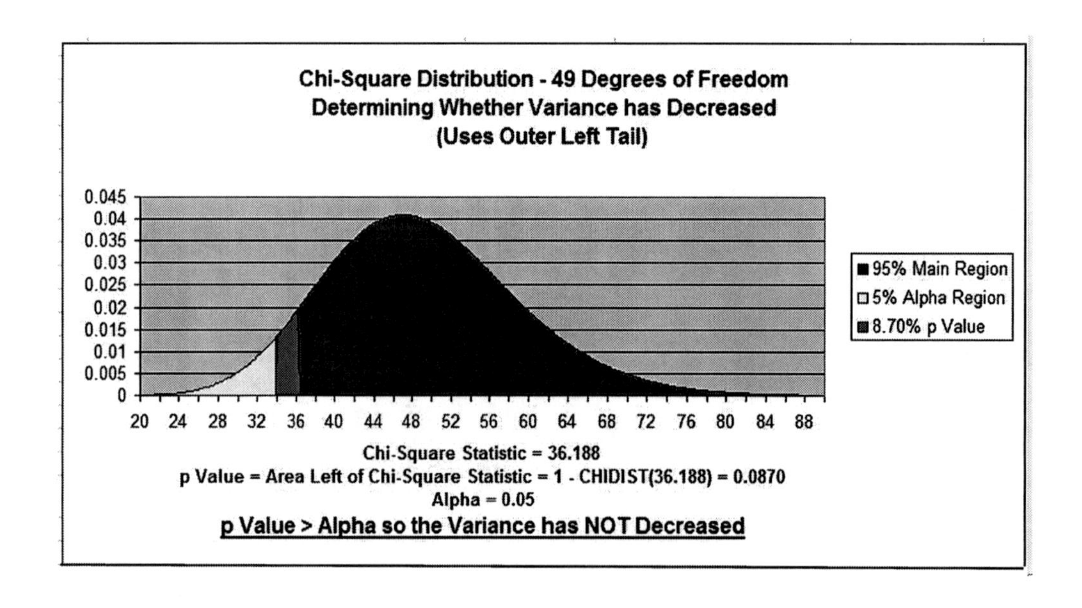

Chapter 15 – Solving Problems With Other Useful Distributions

Multinomial Distribution

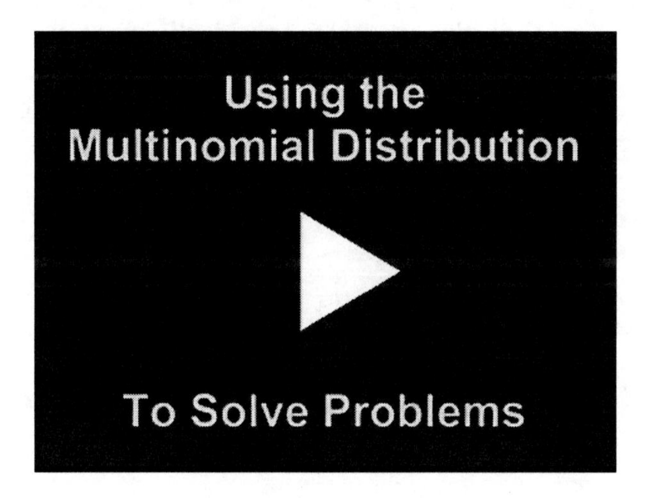

Instructional Video

Go to
http://www.youtube.com/watch?v=fGoEwyvTIkc
to View a
Video About How To Solve
Multinomial Distribution
Problems in Excel

(Is Your Internet Connection and Sound Turned On?)

The Multinomial Distribution is a Discrete distribution, not a Continuous distribution. This means that the objects that form the distribution are whole, individual objects. This distribution curve is not smooth but moves abruptly from one level to the next in increments of whole units.

The Multinomial Distribution provides the probability of a combination of specified outputs for a given number of trials that are totally independent. The probability of each of the individual outputs of each of the trials must be known in order to utilize the Multinomial Distribution to calculate the probability of that unique combination of outputs occurring in the given trials.

Here is the formula for calculating the probability of a multinomial distribution:

$P\,(X1 = n1, X2 = n2, \ldots, Xk = nk) =$

$= [\,(n!) / (n1! * n2! * \ldots * nk!)\,] * [Pr(X1 = n1)]^{n1} * [Pr(X2 = n2)]^{n2} * \ldots * [Pr(Xk = nk)]^{nk}$

The Multinomial Distribution is a generalization of the well-known Binomial Distribution. When k = 2, the Multinomial Distribution is the Binomial Distribution.

An example makes the Multinomial Distribution easier to understand. An example follows:

Problem: A box contains 5 red marbles, 4 white marbles, and 3 blue marbles. A marble is selected at random, its color noted, and then the marble is replaced. 6 marbles are selected in this manner. Find the probability that out of those 6 marbles, 3 are red, 2 are white, and 1 is blue.

Total number of marbles = 12

n = total number of drawings = 6

X_1 = Count of red marbles drawn = n_1 = 3
X_2 = Count of white marbles drawn = n_2 = 2
X_3 = Count of blue marbles drawn = n_3 = 1

The probability of 3 red, 2 white, 1 blue = P(3 red, 2 white, 1 blue)

= $P(X_1 = n_1, X_2 = n_2, ..., X_k = n_k)$
= $P(X_1 = 3, X_2 = 2, X_3 = 3)$

$P(X_1 = n_1, X_2 = n_2, ..., X_k = n_k) =$

$= [(n!) / (n_1! * n_2! * ...*n_k!)] * [Pr(X_1 = n_1)]^{n_1} * [Pr(X_2 = n_2)]^{n_2} * ... * [Pr(X_k = n_k)]^{n_k}$

$= [(6!) / (3! * 2! * 1!)] * [5/12]^3 * [4/12]^2 * [3/12]^1$

= 625 / 5184 = 0.12056 = 12.06%

Excel does not provide the Multinomial Distribution as one of its built-in functions. The user must generate their own Excel formulas. In this case, it would be:

P(3 red, 2 white, 1 blue) =

$= [(6!) / (3! * 2! * 1!] * [5/12]^3 * [4/12]^2 * [3/12]^1$

= **(** FACT(6) / **(** FACT(3) * FACT(2) * FACT(1) **))** * **((** $(5/12)^3$**)** * **(** $(4/12)^2$**)** * **(** $(3/12)^1$**))**

= 0.1206 = 12.06%

Hypergeometric Distribution

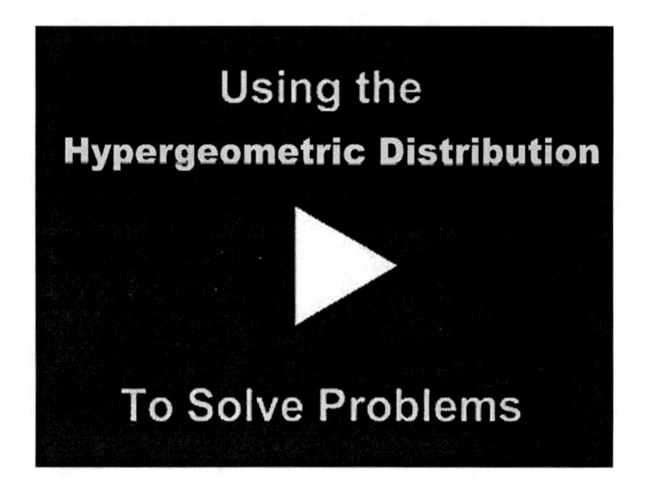

Instructional Video

Go to
http://www.youtube.com/watch?v=57U8vFqVdus
to View a
Video About How To Solve
Hypergeometric Distribution
Problems in Excel

(Is Your Internet Connection and Sound Turned On?)

The Hypergeometric Distribution is almost the same as the Binomial Distribution, except that samples are NOT replaced back into the population. This is known as **Sampling Without Replacement**.

The **Binomial Distribution** calculates the probability of 1 of 2 possible outcomes occurring a certain number of times (x) in a certain number of independent trials (n). The probability of the outcome occurring in a single trial is known (p).

After each trial the samples ARE replaced back into the population when using the Binomial Distribution.

The **Hypergeometric Distribution** calculates the probability of 1 of 2 possible outcomes occurring a certain number of times (x) in a certain number of independent trials (n). The probability of the outcome occurring in a single trial is known (p).

After each trial the samples are NOT replaced back into the population when using the Hypergeometric Distribution.

Excel provides a built-in Hypergeometric Distribution function to calculate the probability of sampling events that are distributed hypergeometrically. This will calculate the probability of one of two possible outcomes occurring a certain number of times in a given number of trials - if the samples are not replaced. Excel's Hypergeometric formula is as follows:

= HYPGEOMDIST (k, n, k_Possible, N)

This function is described on the next page.

HYPGEOMDIST (k, n, k_Possible, N)

Exact number of successes = k

Number of trials = n

Initial possible number of success = k_Possible

Initial population size = N

A graph of the Hypergeometric distribution below is based upon the Hypergeometric problem that follows in which:

Exact number of successes = k = 3

Number of trials = n = 4

Initial possible number of successes = k_Possible ---> Figures in the Horizontal (x) Axis

Initial population size = N = 20

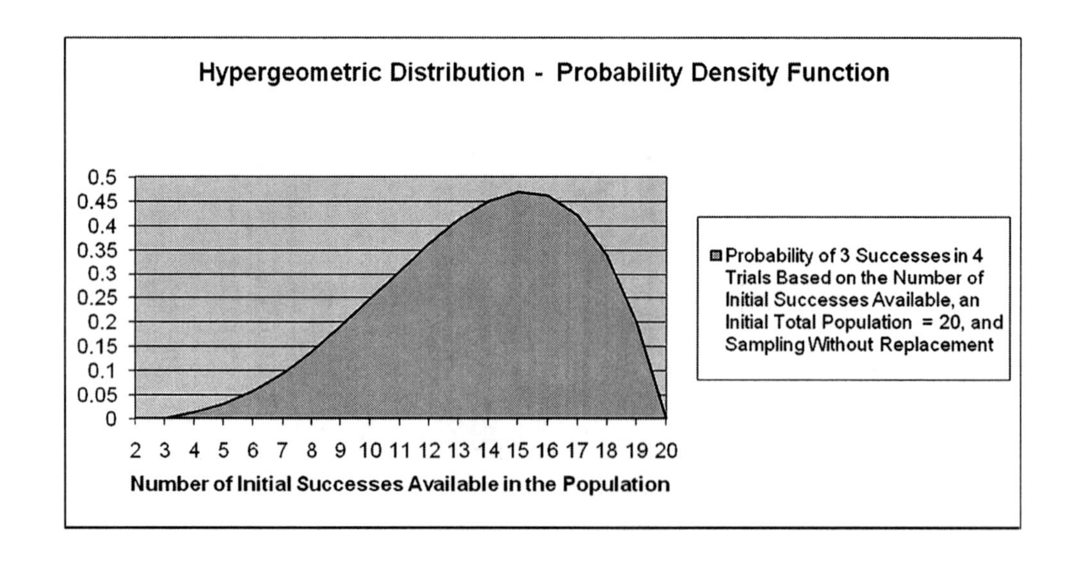

The problem below will illustrate the use of the Hypergeometric Distribution formula:

Problem: A 20-piece chocolate sample consists of 8 caramel samples and 12 nut samples. Calculate the probability that of 4 individual samples 3 taken will produce caramels. Each sample is eaten (not replaced) after it is taken. (If each sample were replaced before the next sample was taken, the Binomial distribution would be used)

Exact number of successes = k = 3

Number of trials = n = 4

Initial possible number of successes = k_Possible = 8

Initial population size = N = 20

= HYPGEOMDIST (k, n, k_Possible, N)

= HYPGEOMDIST (3, 4, 8, 20) = 0.1387 = 13.87%

This point on the graph corresponds to the point on the horizontal axis that equals 8. Note that the y-value of that graph point is 0.1387.

There is a 13.87% probability that 3 out of 4 samples taken without replacement will be caramel samples if the box initially had 20 pieces of candy that included 8 caramel samples.

Sampling WITH Replacement – Using the Binomial Distribution

If this problem were the same except that you were sampling with replacement (putting each candy back instead of eating it before the next sample is taken), you would solve it with the Binomial Distribution as follows:

P (3 Caramels selected in 4 trials with replacement from a box having 20 pieces of candy including 8 caramels) =

= BINOMDIST (3, 4, 8/20, FALSE) = 0.1536 = 15.36%

(Use FALSE to indicate solving not for cumulative function (0 to 3) but solving for exactly 3 caramels chosen in 4 trials with replacement.

Note that the graph at point 3 on the Horizontal has a y value (probability) = 0.1536

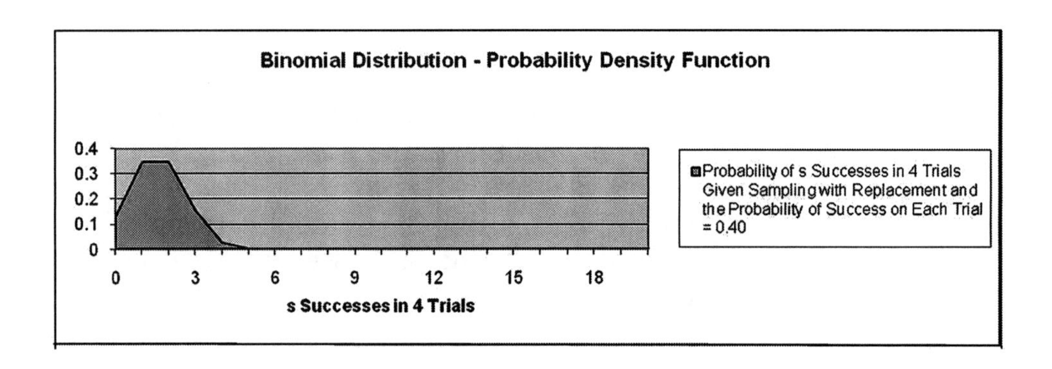

Poisson Distribution

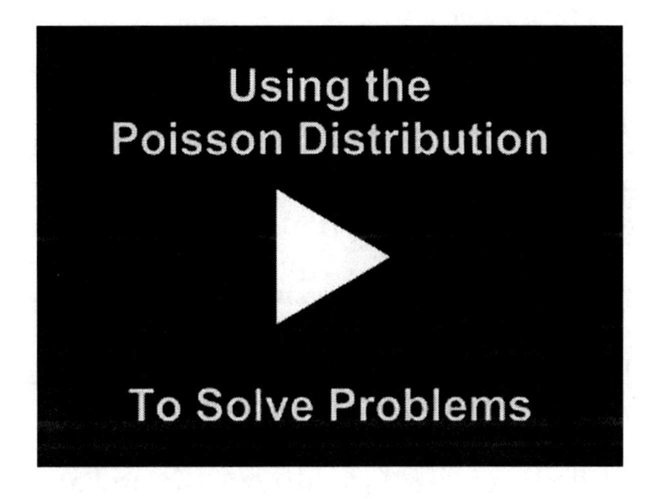

Instructional Video

Go to
http://www.youtube.com/watch?v=bqsfcCSqpBs
to View a
Video About How To Solve
Poisson Distribution
Problems in Excel

(Is Your Internet Connection and Sound Turned On?)

The Poisson Distribution is a widely employed distribution that is used to describe the probability of events that are a result of a rate that occurs over time such as:

Product demand
Demand for services
Number of telephone calls that come over a switchboard
Number of accidents
Number of traffic arrivals
Number of defects

The Poisson Distribution is a Discrete distribution. This means that the events described by this function occur in whole units. The graph of the Poisson Distribution therefore moves from one level to the next in discrete increments, not smoothly.

The Poisson Distribution is used to calculate the probability of a certain number of specific events occurring over a given period of time - if it is known in advance that those events occur in frequency as predicted by the Poisson Distribution. Previous measurement must have been taken to determine: 1) that the events occur in frequency according to the Poisson Distribution, and 2) the average rate, which is the expected number of occurrences of that event over the given time period.

Excel's Poisson formula is given as follows:

POISSON (E(k), k, cumulative?)

= Probability of k or up to k occurrences in a certain time period if the expected number of occurrences in that time period is E(k).

The problem on the next page will better illustrate the use of the Poisson Distribution:

Problem: An average of 4.8 telephone calls per minute is made through the central switchboard according to the Poisson distribution. What is the probability that:

a) exactly 4 phone calls will be made in a given minute

Exact number of events = k = 4

Expected number of events = E(k)

Cumulative Distribution Function? FALSE

We want to calculate the probability of EXACTLY 4 phone calls. This will be the Probability Density Function, not the Cumulative Distribution Function, which would measure the probability of up to 4 phone calls instead of exactly 4.

Pr (Exactly 4 Phone Calls) = Pr (k = 4) =

=POISSON (4, 4.8, FALSE) = 0.182 = 18.2%

Note that the point on the graph that has 4 as the value on the horizontal axis has a value of 0.182 on the vertical axis. The Poisson distribution is a discrete distribution and not a continuous distribution so the graph has corners at each point instead of being smooth.

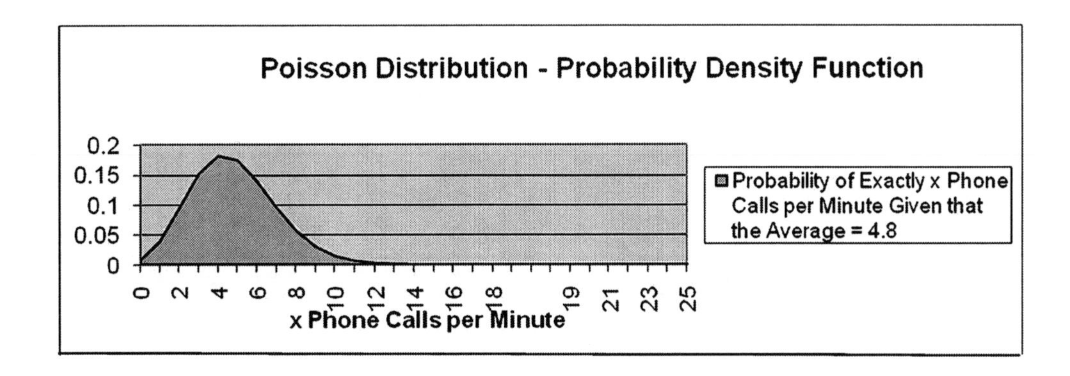

b) up to 4 phone calls will be made in a given minute

Exact number of events = k = 4

Expected number of events = E(k)

Cumulative Distribution Function? TRUE

We want to calculate the probability of UP TO 4 phone calls. This will be the Cumulative Distribution Function.

Pr (Up To 4 Phone Calls) = Pr (k ≤ 4) =

=POISSON (4, 4.8, TRUE) = 0.476 = 47.6%

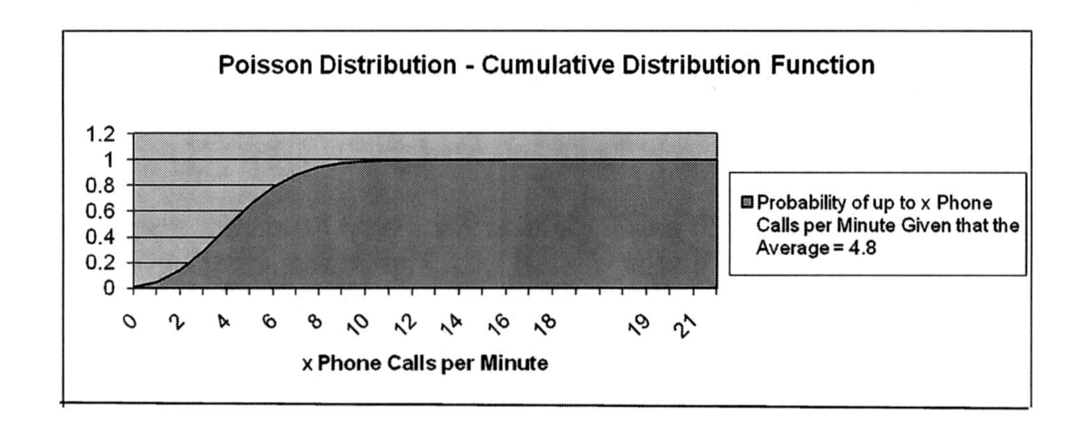

Uniform Distribution

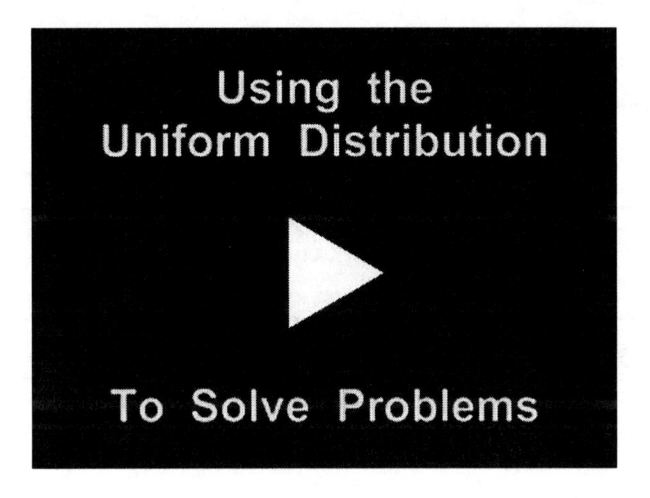

Instructional Video

Go to
http://www.youtube.com/watch?v=U7jS3efn7Mk
to View a
Video About How To Solve
Uniform Distribution
Problems in Excel

(Is Your Internet Connection and Sound Turned On?)

A variable is uniformly distributed if all possible outcomes of that variable have an equal probability of occurring. For example, if a fair die has 6 possible outcomes when rolled once, each outcome has the same 1/6 chance of occurring.

The Uniform Distribution is a Discrete distribution. This means that its events described by this function occur in whole units.

There is no Excel built-in function for the Uniform Distribution. Instead the user must create the Excel calculations. Here is an example:

Problem: A fair die is rolled once. What is the probability that either a 2 or a 5 will appear on top after the roll?

Number of total possible outcomes in 1 trial = 6

Number of times that 2 appears as a possible outcome = 1

Number of times that 5 appears as a possible outcome = 1

Probability of a 2 occurring in 1 roll = 1/6 = 0.1667

Probability of a 2 occurring in 1 roll = 1/6 = 0.1667

Pr (2 occurs) OR Pr (5 occurs) = Pr (2 occurs) + Pr (5 occurs)

= 0.1667 + 0.1667 = 0.333 = 33.33% probability

Exponential Distribution

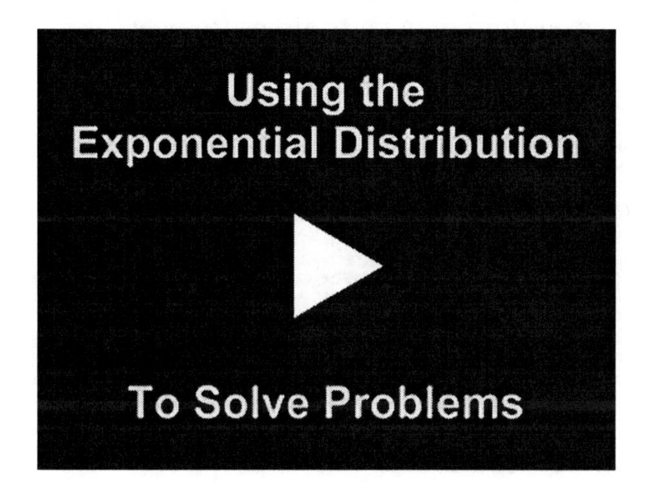

Instructional Video

Go to
http://www.youtube.com/watch?v=E19gxHLpurc
to View a
Video About How To Solve
Exponential Distribution
Problems in Excel

(Is Your Internet Connection and Sound Turned On?)

The Exponential Distribution is used to calculate the probability of occurrence of an event that is the result of a continuous decaying or declining process. The lengths between arrival times in a Poisson process could be described with the Exponential Distribution. Examples of arrival times between Poisson events are as follows:

Time between telephone calls that come over a switchboard
Time between accidents
Time between traffic arrivals
Time between defects

An example of a decaying process that would be predicted by the Exponential Distribution would be:

Time until a radioactive particle decays

The Exponential Distribution is not appropriate for predicting failure rates of devices or lifetimes of organisms because a disproportionately high number of failures occur in the very young and the very old. In these cases, the distribution curve would not be a smooth exponential curve as described by the Exponential Distribution.

The Exponential Distribution predicts time between Poisson events as follows:

Probability of length of time t between Poisson events = $f(t) = ke^{-kt}$
k is sometimes called Lambda

Excel has a built-in Exponential Distribution function as follows:

= EXPONDIST (t, lambda, cumulative?)

Length of time period = t

Decay characteristic = lambda

Cumulative distribution = TRUE or FALSE

(if False, the probability of decay or failure at an exact moment is being calculated using the Probability Density Function)

(if True, the probability of decay or failure UP TO an exact moment is being calculated using the Cumulative Distribution Function)

A problem on the following page will illustrate the use of this function:

Problem: A production machine has a very low defect rate. Time between defects can be predicted by the following Exponential Distribution function:

Time between failures (t) = f(t) = $9\ e^{-9t}$

(t is measured in whole years)

Calculate the probability of a defect being produced within the next 1/10th year.

(Using "Within" indicates that the Cumulative Distribution function will be used)

t = 1/10 = 0.10

k = Lambda = 9

Calculate Cumulative = TRUE Distribution function?

Probability of defect occurring in 1/10 years

=EXPONDIST(0.1,9,TRUE) = 0.5934 = 59.34%

Note that the graph point at Time t = 0.1 has the probability of 0.5934.

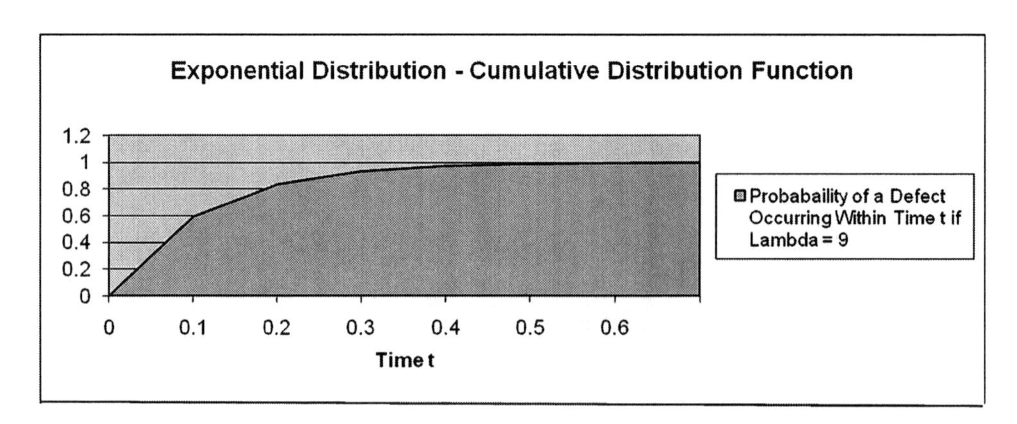

Gamma Distribution

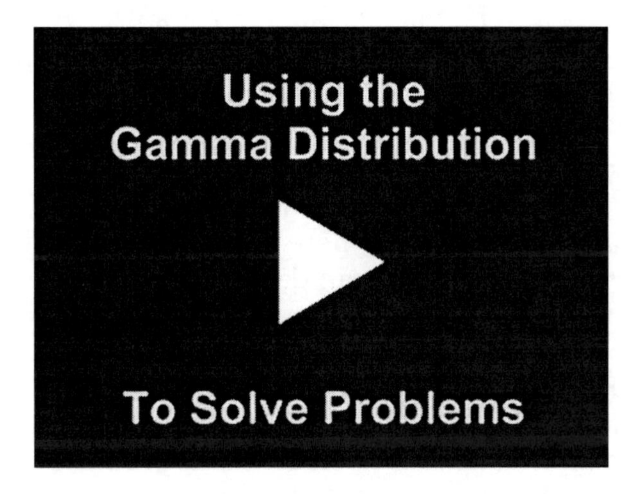

Instructional Video

Go to
http://www.youtube.com/watch?v=-w-CukEFT9g
to View a
Video About How To Solve
Gamma Distribution
Problems in Excel

(Is Your Internet Connection and Sound Turned On?)

The Gamma Distribution represents the sum of n exponentially distributed random variables. Applications of the Gamma Distribution are often based on intervals between Poisson-distributed events. Examples of these would include queuing models, the flow of items through manufacturing and distribution processes, and the load on web servers and many forms of telecom.

Due to its moderately skewed profile, it can be used as a model in a range of disciplines, including climatology where it is a working model for rainfall, and financial services where it has been used for modeling insurance claims and the size of loan defaults. It has therefore been used in probability of ruin and value-at-risk equations.

The Gamma Distribution function is characterized by 2 variables, its shape parameter alpha (α) and its scale parameter theta (Φ). The Gamma Distribution function calculates the probability of wait time between Poisson distributed events to be time t,

Excel has a built-in Gamma function as follows:

GAMMA (t Units of Time, α, Φ, Cumulative?)

= the probability that a Poisson-distributed event will occur at or within t Units of Time if the event can be characterized with a specific α and Φ.

Units of time = t

Alpha = α = shape parameter

Theta = Φ = Scale parameter

Cumulative? = if TRUE, calculating probability of event occurring within t units of time. If FALSE, calculating probability of event occurring at exactly t units of time.

Problem: Calculate the probability of the a Poisson-distributed event occurring before Time t = 10 if the Gamma Distribution function has alpha, α, = 2 and theta, Φ, = 4.

Units of waiting time until event occurs = t = 10

Alpha, α = 2

Theta, Φ = 4

Cumulative Distribution Function? = TRUE

Probability of this event occurring within 10 units of time (cumulative function)

= GAMMADIST (10, 2, 4, TRUE) = 0.7127 = 71.27%

Note that the graph point at Time t = 10 has a probability of 0.7127.

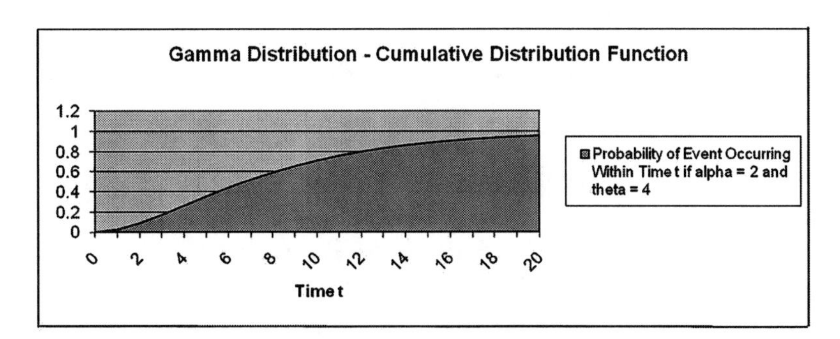

The Chi-Squared Distribution is a Gamma distribution in which the shape parameter, α, is set to the degrees of freedom divided by 2 and the scale parameter, theta, is set to 2.

The Gamma Distribution with its shape parameter, α, set to 1 and its scale parameter, theta, set to b, become the Exponential Distribution with k, lambda, set to b.

Beta Distribution

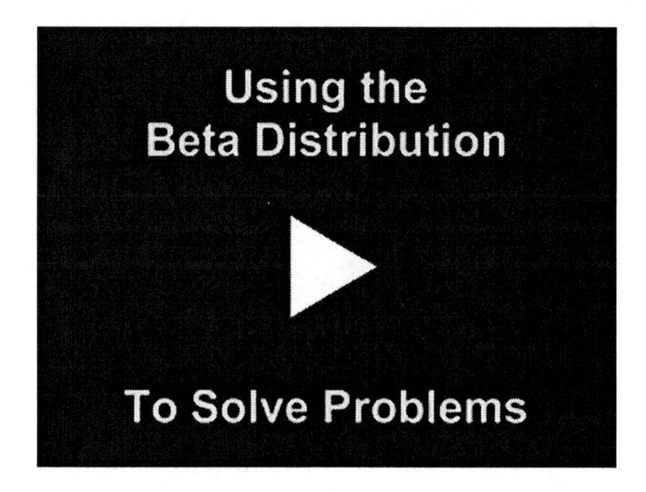

Instructional Video

Go to
http://www.youtube.com/watch?v=aZjUTx-EoPk
to View a
Video About How To Solve
Beta Distribution
Problems in Excel

(Is Your Internet Connection and Sound Turned On?)

The Beta Distribution models events which are constrained to take place between a minimum and maximum time limit. For this reason, the Beta Distribution is often used for modeling project planning and control systems such as PERT (Project Evaluation and Review Technique) and CPM (Critical Path Method). The Beta Distribution is often used to calculate the probability that a project will be completed within a given period of time. Below is an example which illustrates its use:

Problem: Calculate the probability of completing the following project before Time t = 5 if the project is described by the following parameters:

Evaluation time period = t = 5

Alpha, α, = 8
Beta, β, = 10
Minimum completion time in units of time = 2
Maximum completion time in units of time = 7

Probability of completing the task within time = 5 units of time (within = cumulative function)

= BETADIST (t, α, β, Min completion time, Max completion time)
= BETADIST (5, 8, 10, 2, 7) = 0.908
= 90.81%

Note that the graph point at Time t = 5 has a probability of 0.908.

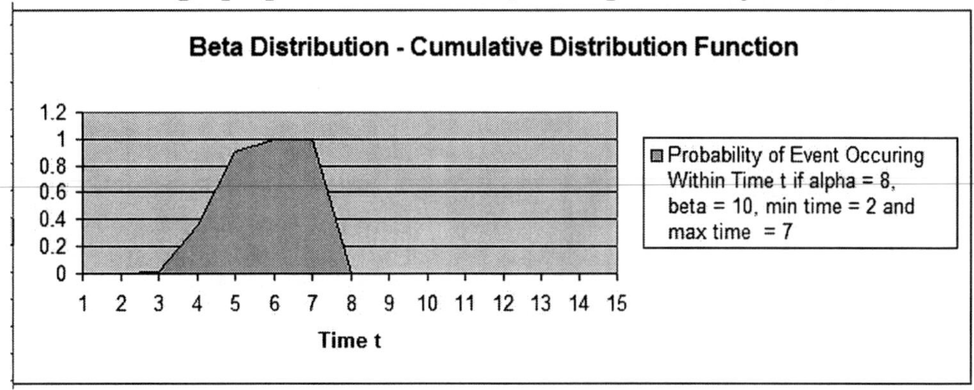

Weibull Distribution

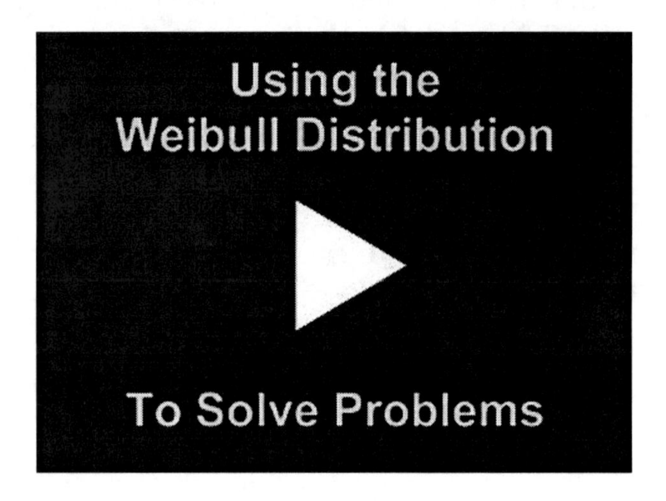

Instructional Video

Go to
http://www.youtube.com/watch?v=g4FHaBhecTM
to View a
Video About How To Solve
Weibull Distribution
Problems in Excel

(Is Your Internet Connection and Sound Turned On?)

The Weibull Distribution is a special case of the Generalized Extreme Value distribution. The Weibull distribution has been used extensively as a model of time to failure for manufactured items and has become one of the principal tools of reliability engineering. The applications of the Weibull Distribution have expanded and include Finance and Climatology. There are three parameters of the Weibull distribution: time t, α - alpha (the shape parameter), and β (the scale parameter).

$\alpha > 1$ --> Failure rate increases over time (suggests "wear out")
$\alpha = 1$ --> Constant failure rate - Items fail from random events
$\alpha < 1$ --> Failure rate decreases over time (suggest high "infant mortality")

The problem on the next page illustrate the use of how this function can be used to solve a problem.

Problem: Calculate the probability that a part will fail at time = 2 if the part's failure occurrence is Weibull-distributed and has α = 0.5 and β = 4.

t = Time = 2

α = Alpha = 0.5

β = Beta = 4

We are determining the probability of part failure at exactly time t = 2 so we are using the Probability Density Function.

Probability of part failure occurring at exactly Time t = 2 given that time to part failure is Weibull-distributed with α = 0.5 and β = 4 is calculated is follows:

= WEIBULL(2,0.5,4,FALSE) = 0.087 --> 8.7% probability

Note that the graph point at Time t = 2.0 has a probability of 0.087.

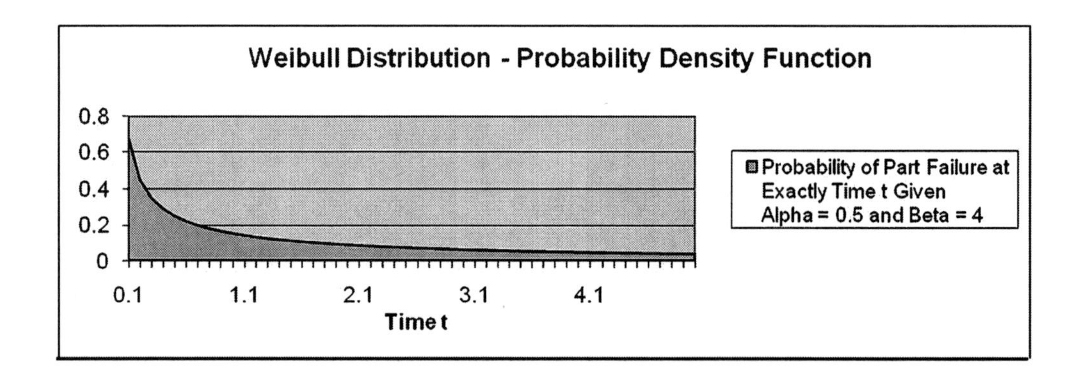

Problem: Calculate the probability the a part will fail by time = 2 if the part's failure occurrence is Weibull distributed and has α = 0.5 and β = 4.

Time t = 2

α Alpha = 0.5

β = Beta = 4

We are determining the probability of part failure at exactly time t = 2 so we are using the Probability Density Function.

Probability of part failure occurring at exactly Time t = 2 given that time to part failure is Weibull-distributed with α = 0.5 and β = 4 is calculated is follows:

= WEIBULL(2,0.5,4,TRUE) = 0.506

--> 50.6% probability

Note that the graph point at Time t = 2 has a probability of 0.506.

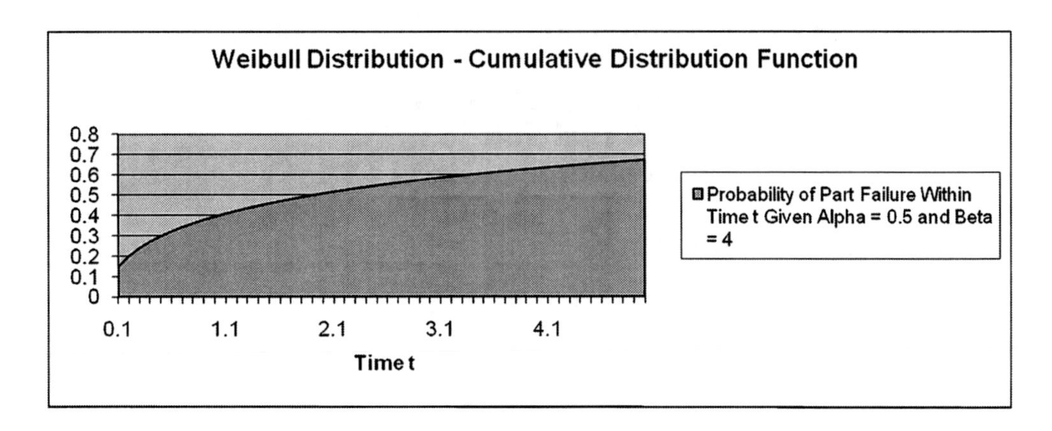

F – Distribution

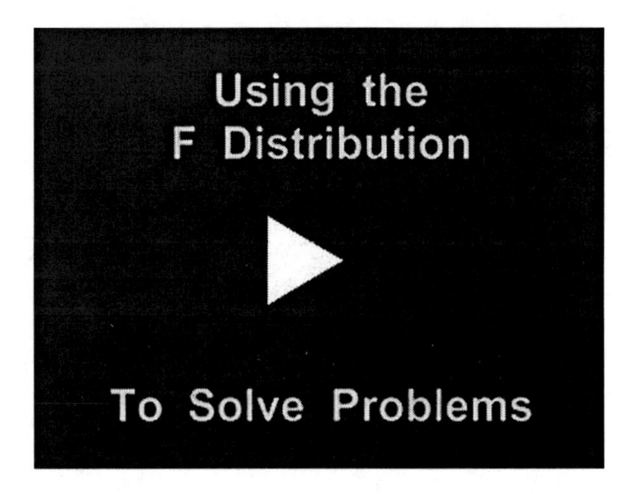

Instructional Video

Go to
http://www.youtube.com/watch?v=y2NLn5DDwko
to View a
Video About How To Solve
F Distribution
Problems in Excel

(Is Your Internet Connection and Sound Turned On?)

The F Distribution is used to determine whether two groups have different variances. The F Distribution is normally used to develop confidence intervals and hypothesis tests. It is rarely used for modeling applications.

The F Distribution has 4 parameters: X^2_1 (the calculated Chi-Square statistic for data group 1), X^2_2 (the calculated Chi-Square statistic for data group 2), v_1 (the degrees of freedom of group 1), and v_2 (the degrees of freedom of group 2). An example of how the Chi-Square statistic is calculated from a group of data can be found in the course module entitled "Chi-Square Independence Test."

The F Distribution is actually a family of distributions. Each different F Distribution has a unique combination of v_1 and v_2.

An individual F Distribution is actually the distribution of the F Statistic. The formula for the F Statistic is as follows:

F Statistic = $(X^2_1 / v_1) / (X^2_2 / v_2)$

As stated, the F Distribution is rarely used for modeling applications, but is often used for developing confidence intervals and hypothesis tests. Because of this, the most important use of a particular F Statistic is the calculation of its p Value. The p Value equals the percentage of total area under that unique F Distribution curve to the right of the given F statistic (and therefore the area in the outer curve tail to the right of the F Statistic). The Excel formula for the p Value for a particular F Statistic within its unique F Distribution is:

p Value = FDIST = (F Statistic, v_1, v_2)

The p Value is compared with α, the required Level of Significance. If the p Value is less than α, then the two data groups are assumed to have different variances. If the p Value is greater than α, the two data groups are assumed to have equal variances.

ANOVA (Analysis of Variance) tests calculate an F Statistic and its corresponding p Value for each pair of data sets that are being tested for independence. If the p Value pertaining to a a pair of data sets is less than the required α (which, for example, would equal 0.05 if a 95% Level of Certainty was required), it is assumed that the pair of data sets is not independent of each other. Refer to the ANOVA module of this course to see examples of this.

Calculating the F Statistic between two data sets involves a lot of work. A complete example of the calculation of the F Statistic and p Value between two data sets is shown at the end of the ANOVA module. Here, a hand-calculation of the F Statistic and p Value is performed to determine if there is a relationship between sales closing methods and sales results. The problem required a 95% Level of Certainty. The α (Level of Significance) was therefore equal to 0.05. The p Value between the two data sets was calculated to be 0.144. This p Value is less than α so sales are assumed to be related to the closing method used. Sales results and closing methods are assumed to not be independent of each other because different closing methods are shown to produce different sales results.

A summary of that problem is as follows:

The problem requires determination of whether closing methods used have an affect on sales. Three sales groups were each required to use a different closing method for the entire test. The total sales results from each group were recorded. ANOVA analysis was employed to determine with a 95% Level of Certainty whether the choice of closing method affected the level of sales.

The ANOVA process breaks the data down in two ways for analysis. One grouping of data is labeled the "Between Groups" data. The other grouping of the data is labeled the "Within Groups" data.

An F Statistic and its subsequent p Value were calculated based upon these two groups of data. The p Value (0.0144) was found to be less than α (0.05 --> based upon the 95% Level of Certainty). Therefore this implies that sales are not independent of closing method used.

A summary of the calculations is as follows:

"Between Groups" data grouping:

X^2_1 = Chi-Square Statistic$_{\textbf{Group1}}$ = 72

v_1 = degrees of freedom$_{\textbf{Group1}}$ = 2

"Within Groups" data grouping:

- X^2_2 = Chi-Square Statistic$_{\textbf{Group 2}}$ = 46

- v_2 = degrees of freedom$_{\textbf{Group 2}}$ = 9

F Statistic = (X^2_1 / v_1) / (X^2_2 / v_2)

= (72 / 2) / (46 / 9)

= 7.043478261

p Value = FDIST (F Statistic, v_1, v_2)

= FDIST (7.043478261, 2, 9)

= 0.0144

0.0144 is less than **α** (0.05) so it is assumed that the two groups are not independent. Sales are therefore related to the closing method used because the variances are different.

Chapter 16 - How To Graph Distributions

Graphing Overview

Overall, this course module will demonstrate how to create a chart from generic data. You have complete understanding of a creating chart from generic data before moving on to charting statistic distributions like the Normal or Chi-Square distributions.

The basic concepts of chart creation remain the same with all charts. The mean difference with creating charts of statistical distributions is the input data. All charts will be graphed in the same way except that different data sets will be plugged into each different chart. For this reason, it is very important to master graphing a chart with generic data before moving on to graph statistical distributions in the second half of this course module.

Incidentally, all graphs in this entire course using the instructions given below. With that, let's learn to graph a chart with generic data:

1) Learning how to graph a generic set of x - y coordinates.

Here is a set of x - y coordinates that will be graphed:

x	y
1	2
2	4
3	7
4	10
5	15

To begin the process of creating a chart in Excel, go into Excel and do the following:

Insert / Chart

Then select the type of chart shown in the diagram below and click Next:

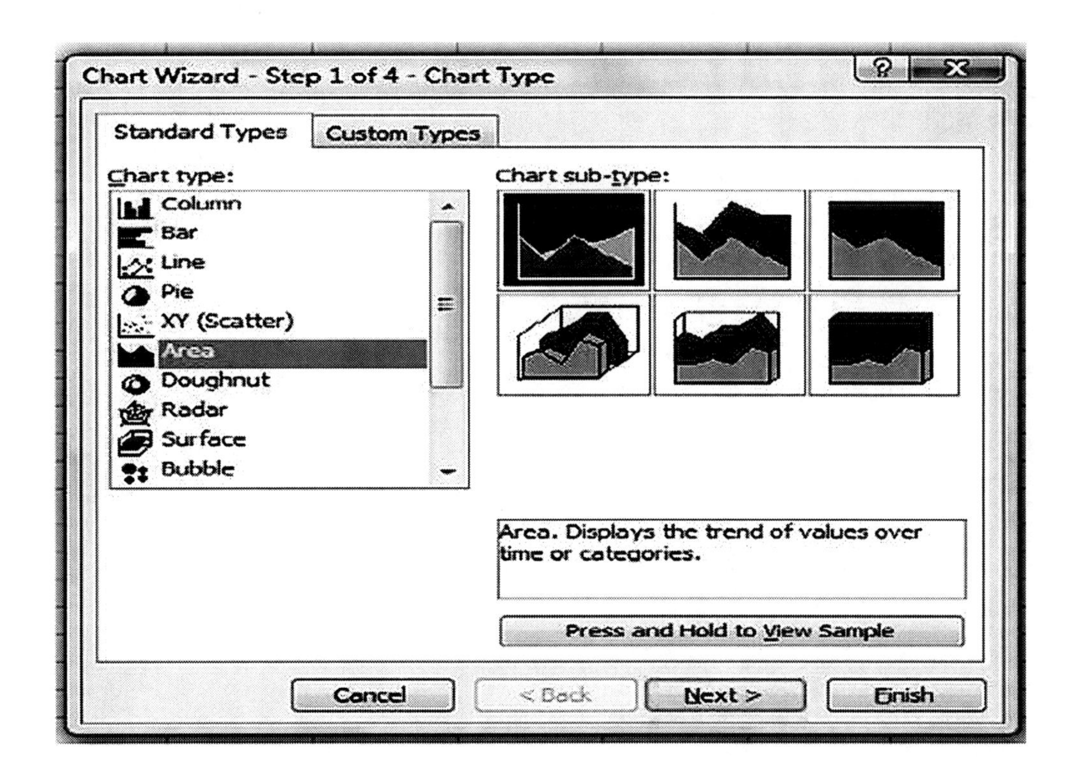

Click on Columns and then click the Series tab:

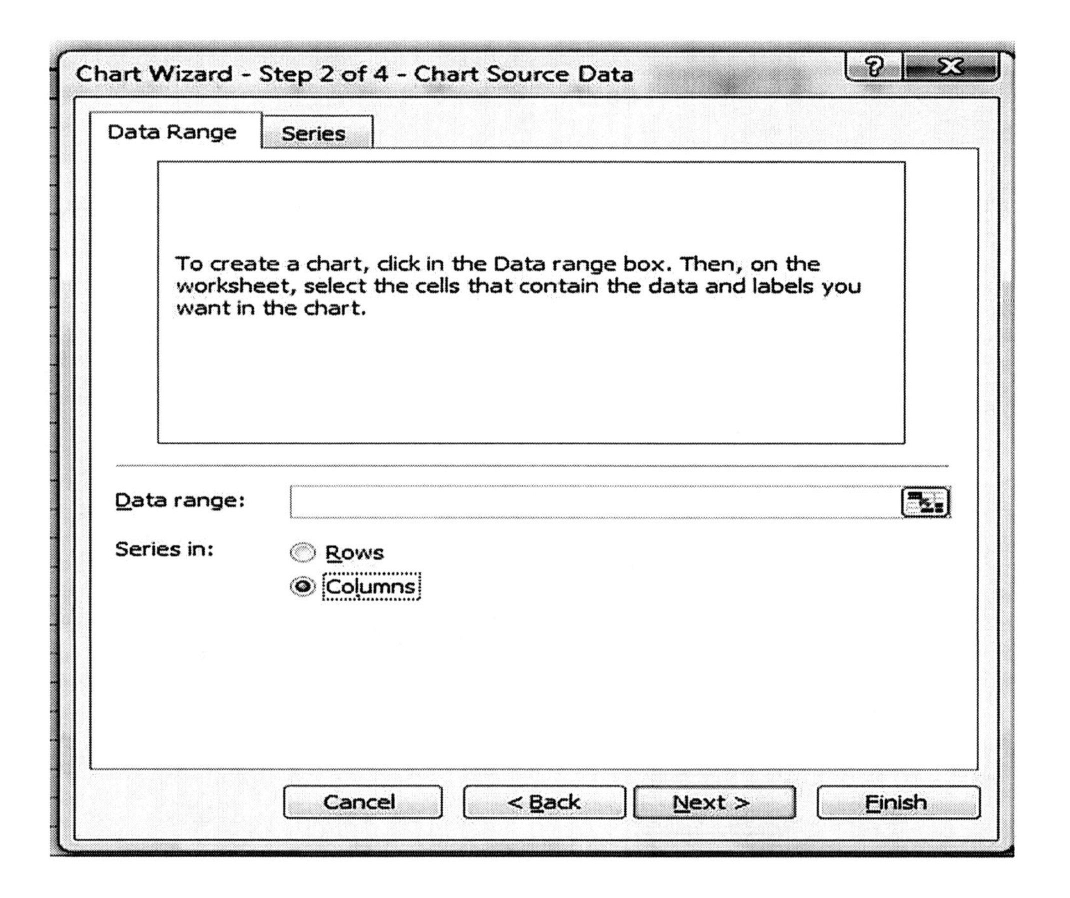

Click the Add Button to add a series of y-coordinate data:

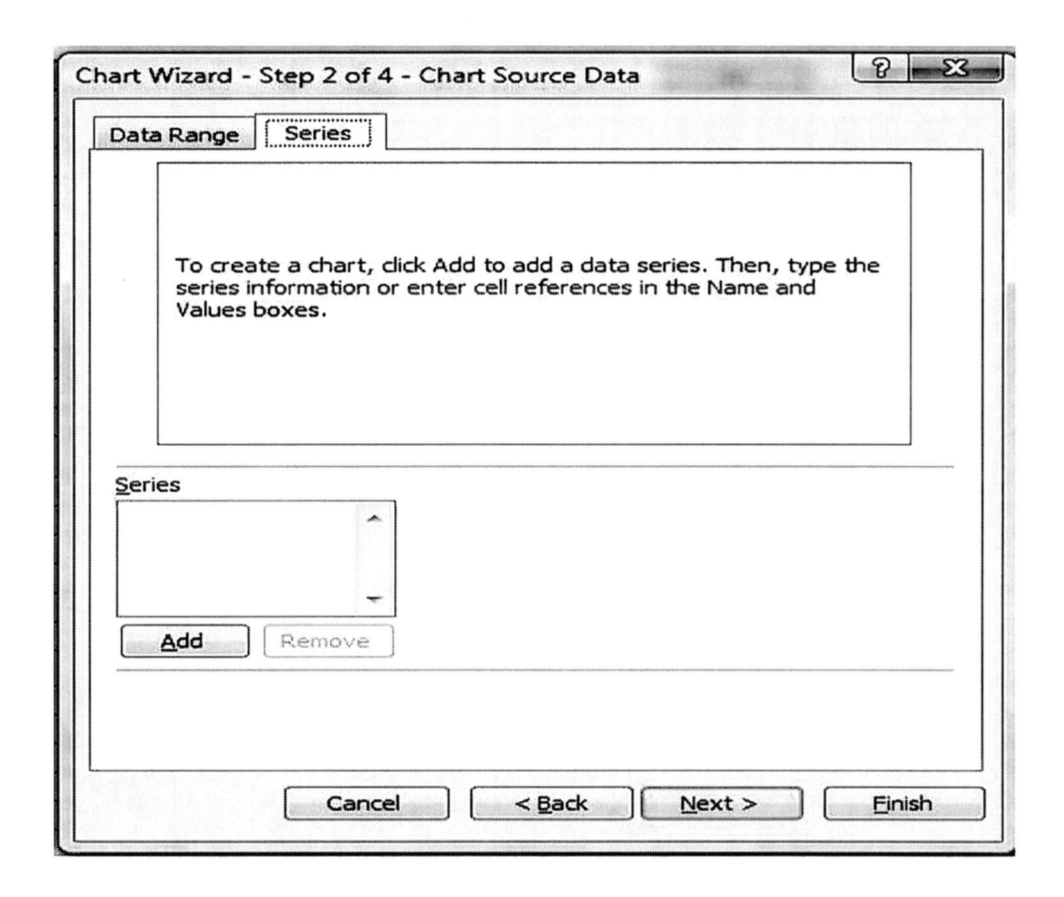

Highlight the default data in the Values input box. This is done below and everything in the Values input box is now highlighted dark black:

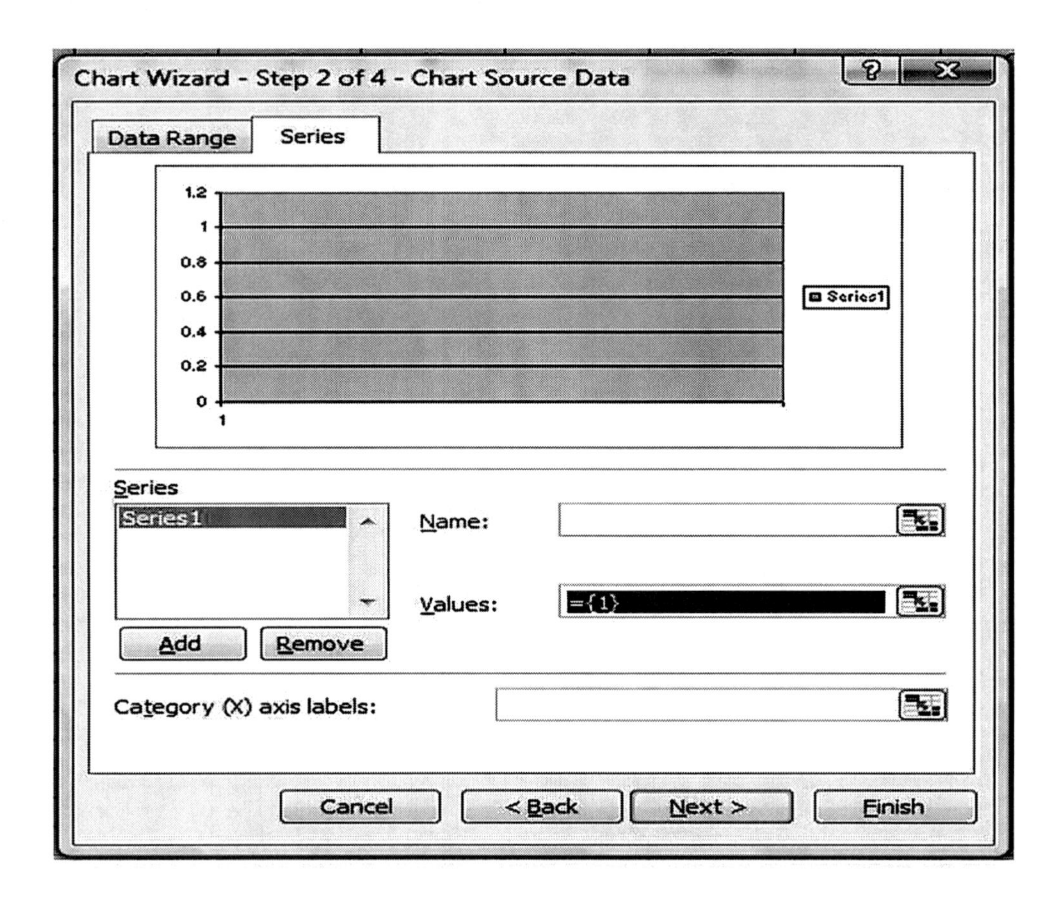

Highlight the y coordinates of what will be graphed as is highlighted in yellow here:

x	y
1	2
2	4
3	7
4	10
5	15

Highlighting the y-coordinates will cause their spreadsheet location to be recorded in the Values input box. In this case, the Y coordinates were labeled "Generic Y Coordinates." What you are seeing in the Values input box (=”How to Graph Distributions) is part of the address of the data. This is name of the spreadsheet that held the data. The cell addresses follow the name of the spreadsheet and are part of that whole address. A graph of the y data will be shown on the next page as well:

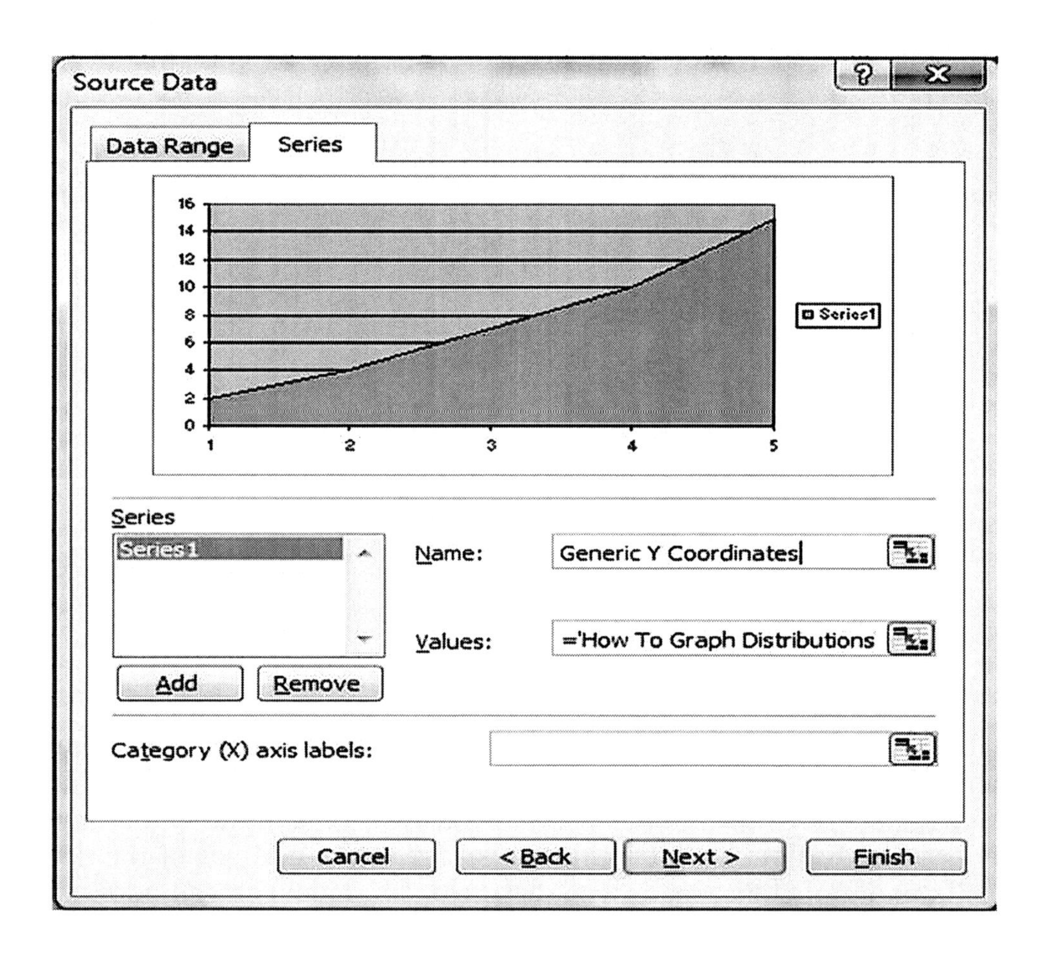
Source Data
Data Range
Series
16
14
12
10
8
6
4
2
0
1
2
3
4
5
Series1
Series
Series1
Name:
Generic Y Coordinates
Values:
='How To Graph Distributions
Add
Remove
Category (X) axis labels:
Cancel
< Back
Next >
Finish

Now, click the cursor in the Category (X) axis labels and then highlight the x coordinate data:

x	y
1	2
2	4
3	7
4	10
5	15

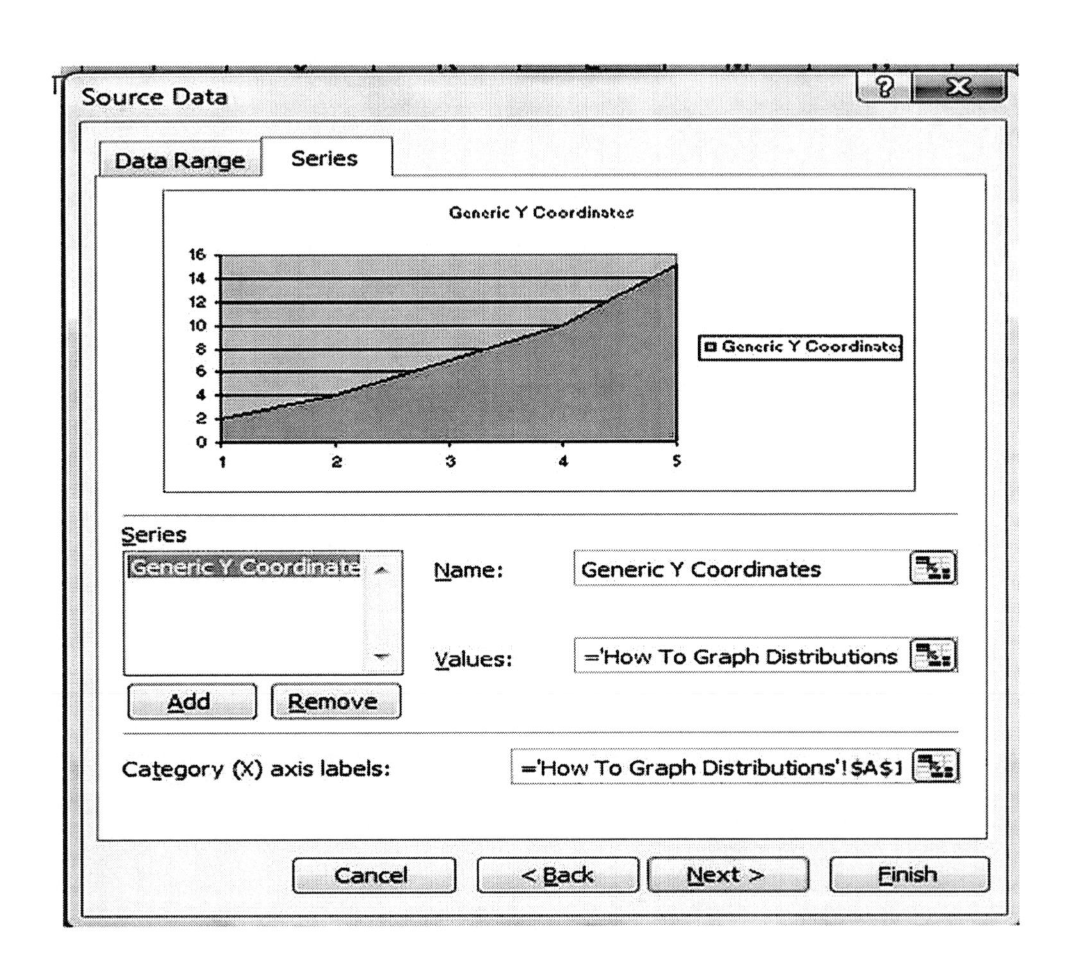

Select the Titles Tab:

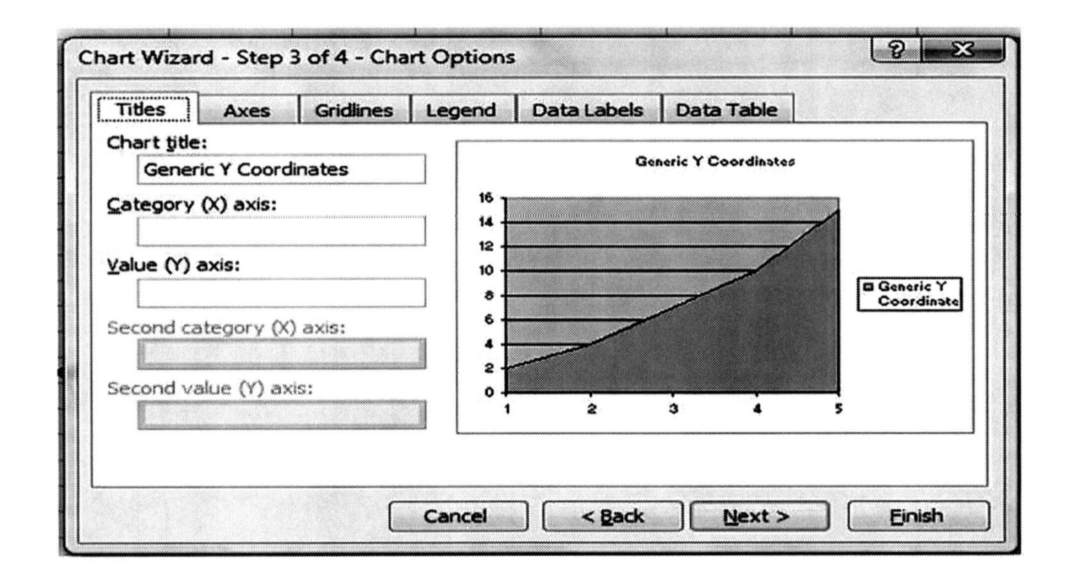

Type in a Chart Title, Category (X) axis, and Value (Y) axis and click Next:

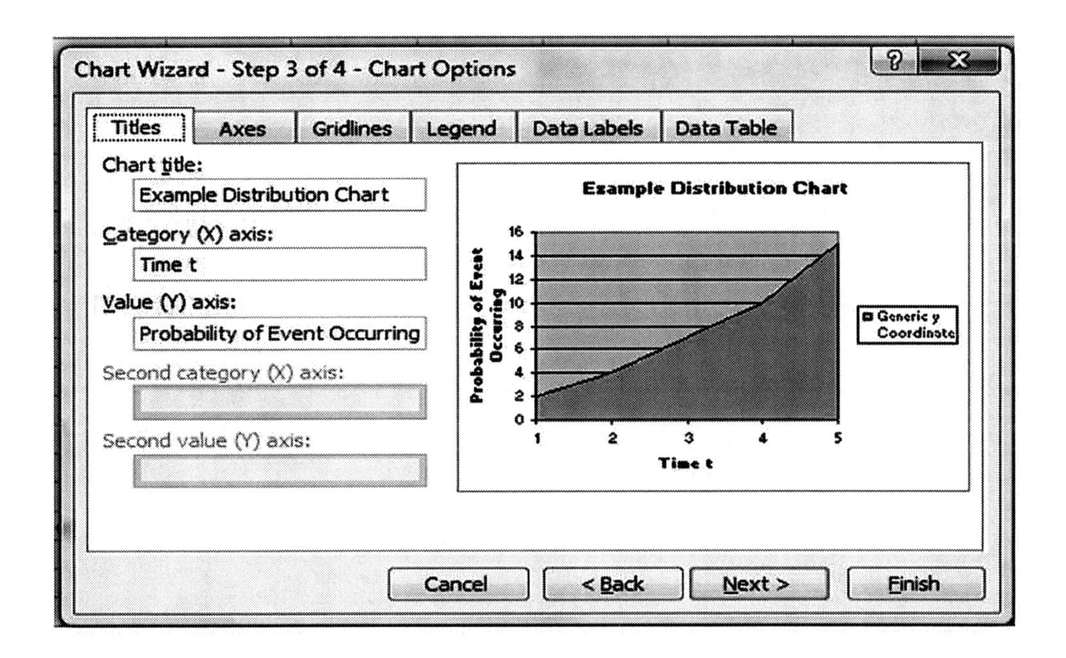

Click Finish in the dialogue box below.

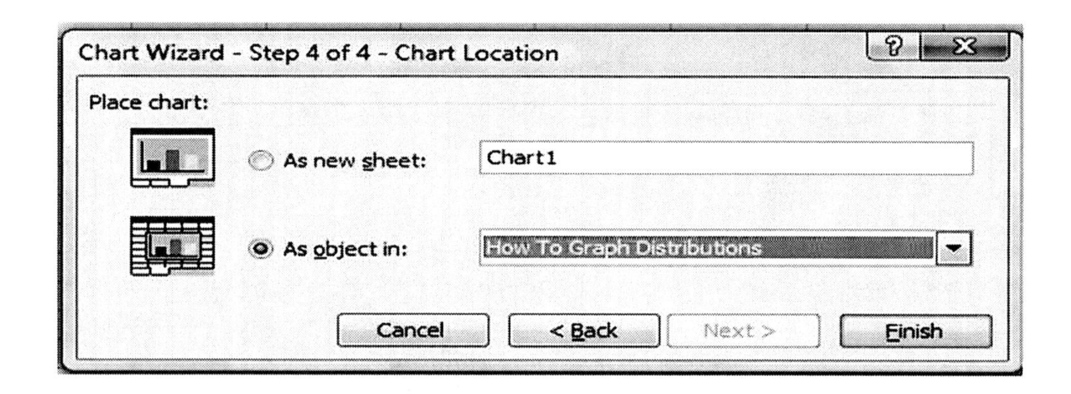

The resulting graph appears:

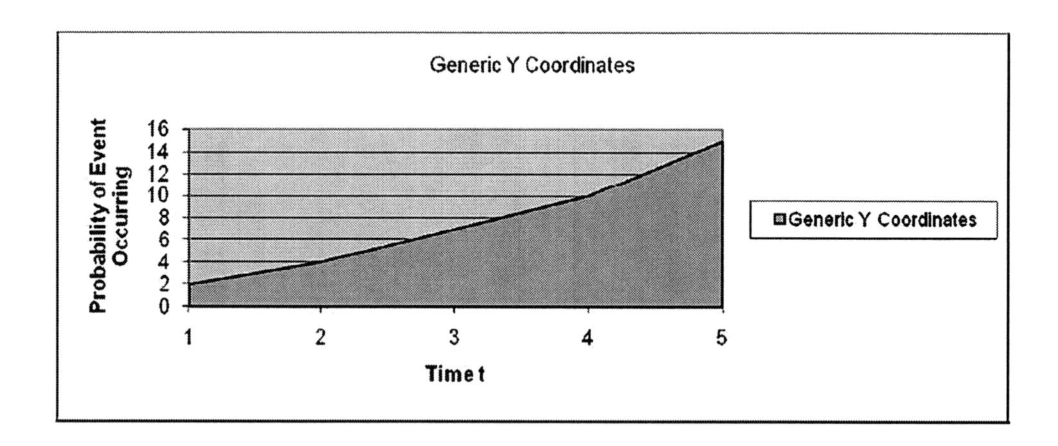

2) Learning how to create the x coordinates and the y coordinates specific to the type of distribution being graphed.

Before you can begin to graph specific statistical functions, you must have mastered creating graphs of generic data. This was demonstrated in the first half of this course module. If you are not very familiar with the process of graphing generic data, please review the first half of this course module.

Creating the graphs of statistical functions is exactly the same as creating graphs using generic data. The main difference is that different sets of data are used for the Series 1 and for the (x) axis labels. Each of these sets of data is provided for each statistic function to be graphed.

Each statistical function graph shown below is a fully functioning graph copied directly from its Excel spreadsheet. The yellow cells are user inputs. Any changes made to the contents of any of these cells will be reflected in its appropriate graph immediately.

If you reconstruct the graphs in Excel exactly as displayed below, you should produce the same result – a fully functioning graph of that statistical function that changes its shape according to the user inputs typed into the yellow user input cells.

Each of the graphs has its own vertical and horizontal Excel address bars so that you will be able to replicate the formulas into the same corresponding cells in the Excel spreadsheet on your computer. For example, if a formula is typed into cell D4 in an example below, you should type the same formula into the cell D4 on the Excel spreadsheet on your computer.

The blue-colored cells below hold formulas that should be "copied down" the column. The exact formula to be copied into each specific blue cell will be displayed nearby in a tan-colored rectangular area. After you type the exact formula as it is shown into the indicated blue cell on your spreadsheet, you must then "copy the cell down." This is a process which copies the contents of the initial cell into each cell below. To "copy the cell down," mouse over the lower right corner of the initial cell. As you mouse of the lower right corner of this cell, you will notice that the cursor changes shape to a small cross. When that happens, right-click and "drag" the contents of the cell down the column while continuing to hold the mouse down (keeping the right-click button depressed).

Drag the formulas down the same number of cells as is done in the appropriate example below. When you release the mouse, you will notice that all of the cell that you "dragged" over now contain the formula and the correct result of the formula. The formulas below have been constructed so that you don't need to worry about relative and absolute addresses. If you are not familiar with those concepts, don't worry about it. Copying the formula from the top cell that you have typed it into down to the bottom of the column should produce the correct formula in each successive cell.

When you are constructing the graphs with Excel, you will need to know which sets of column data to specify as (x) axis labels and as Series 1 data. Each statistical function example below states over the top of the data columns whether it is to be used as Series 1 data. Only one example - Graphing the Outer Tails of the Normal Distribution – requires Series 2 Data. This is clearly labeled in the example.

The data column for each example that should be used as the (x) axis labels should be fairly apparent in each example. If it is not, simply look at the completed graph for each example and use the data column that is labeled with the same label as the x-axis on the completed graph.

When you have completed inputting the correct data columns into Excel and you are ready to view the output of the graph, you may designate that the output graph appears in exactly the same place on your spreadsheet as it appears in the example spreadsheet below.

You may find the process of graphing statistical functions a little confusing at first, but as soon as you complete your first one, you will be able to graph all of the other functions fairly easily. Just reconstruct everything exactly as it appears - the same formulas in the same cells.

You'll notice that the t-Distribution and the Chi-Square distribution graphs had to be constructed from the actual formulas instead of using built-in Excel functions. This is the result of the built-in function producing an incorrect result (the t-Distribution) or no built-in function existing (the Chi-Square Distribution). For these distributions, you must take great care that you copy the formulas exactly as they appear in the examples and in exactly the same cells. Any mistake of a single character or cell address will cause the graphing function to produce an incorrect result. This is actually the case with all of the distributions, but the t-Distribution and Chi-Square Distribution are much more computationally intensive because we must deal with the formulas and not built-in Excel functions.

Most distributions have two major functions that can be graphed. These two functions are the Probability Density function and Cumulative Distribution function.

The **Probability Density Function** provides the probability that the distributed variable will assume a certain value. For example, a Probability Density function might calculate the probability of an event occurring at exactly time = 5. The range of possible probabilities is listed on the vertical or y-axis.

The **Cumulative Distribution Function** provides the probability that the distributed variable will assume any value up to a certain point. For example, a Cumulative Distribution function might calculate the probability of an event occurring at any time up until time = 5. The Cumulative Distribution function would equal the sum of the probabilities calculated by the Probability Density function at all points until time = 5. The Cumulative Distribution function values are shown on the vertical or y-axis and range in values from 0 to 1, which corresponds to 0% to 100%.

Below are displayed the x coordinate data, the y coordinate data, and the completed graph for the Probability Density function and / or the Cumulative Distribution function for most of the distributions mentioned in this course.

To build any of the graphs, substitute the x and y-coordinate data for the generic data and follow the same chart-building instructions.

Some of the distributions have built-in Excel functions which calculate the Probability Density function and / or the Cumulative Distribution function. Some do not. In some cases in which Excel did not provide a built-in function, a hand-calculation of the Probability Density function and / or the Cumulative Distribution function was used.

Normal Distribution

Probability Density Function

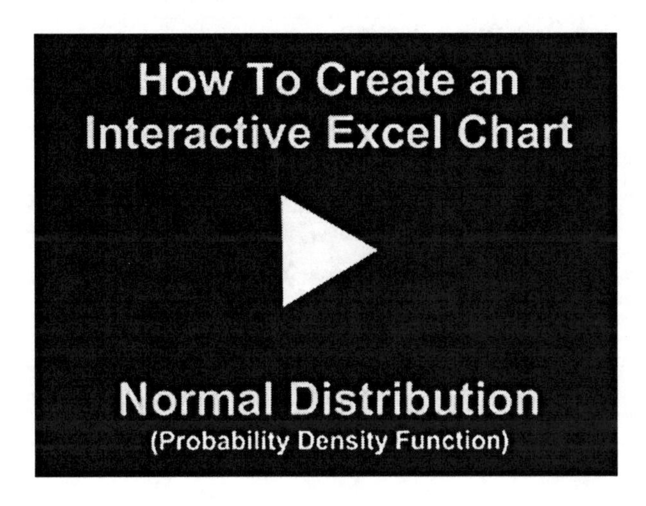

Instructional Video

Go to
http://www.youtube.com/watch?v=h4Zd7iGB4aI
to View a
Video About How To Create
a User-Interactive Graph of
the Normal Distribution's
Probability Density Function
in Excel
(Is Your Internet Connection and Sound Turned On?)

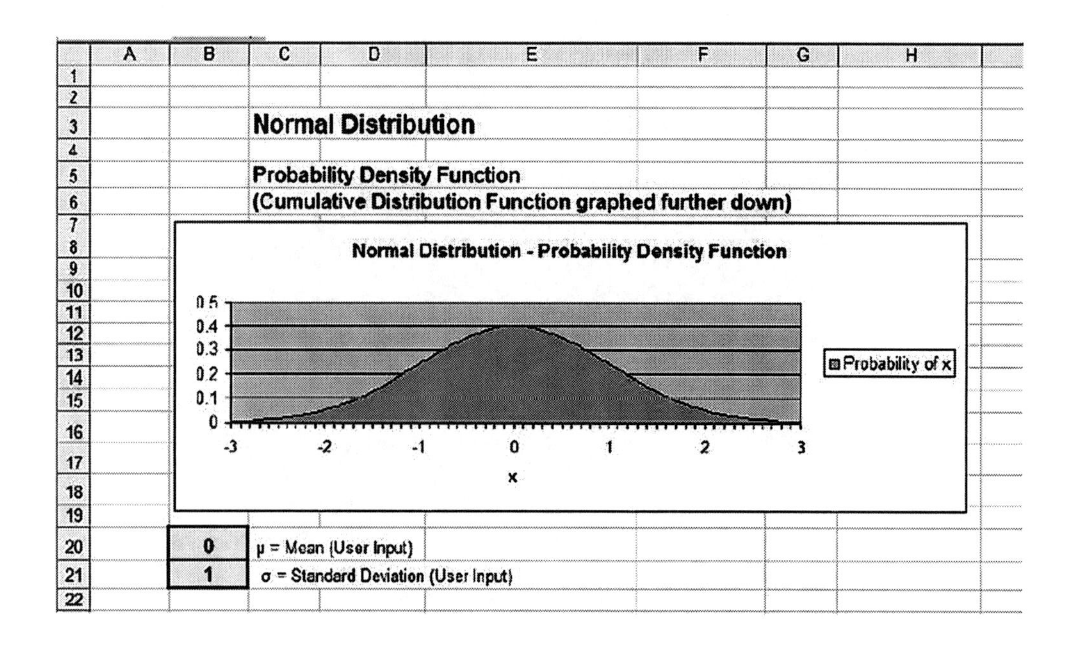

Normal Distribution
Probability Density Function
(Cumulative Distribution Function graphed further down)
Normal Distribution - Probability Density Function
0.5
0.4
0.3
0.2
0.1
0
-3
-2
-1
0
1
2
3
x
Probability of x
0
μ = Mean (User Input)
1
σ = Standard Deviation (User Input)

	A	B	C	D	E
19					
20		0	μ = Mean (User Input)		
21		1	σ = Standard Deviation (User Input)		
22					
23					
24		A specific Normal Curve can be completely constructed if only the			
25		mean and standard deviation are known. These are the two			
26		user inputs for this curve. Each formula to the right of a cell that			
27		is shown below is the formula within that cell. That formula is			
28		copied to end of each column of data. The generic chart-building			
29		process is applied to this data to construct the graph below the data.			
30					
31		To create the columns of data below which are the inputs for the graph			
32		and change when new data is typed into the user inputs above, copy the			
33		given formulas below into the designated cells, then copy the formula			
34		down the column by simply dragging the formula down the column:			
35					
36		C46 = B46*$B$21/10			
37					
38		D46 = C46+$B$20			
39					
40		E46 = NORMDIST(D46,$B$20,$B$21,FALSE)			
41					
42					
43					Probability
44					Density
45				X Axis	Function
46		-30	-3	-3	0.004431848
47		-29	-2.9	-2.9	0.005952532
48		-28	-2.8	-2.8	0.007915452
49		-27	-2.7	-2.7	0.010420935

	A	B	C	D	E	F
43					**Probability**	
44					**Density**	
45				**x Axis**	**Function**	
46		-30	-3	-3	0.004431848	
47		-29	-2.9	-2.9	0.005952532	
48		-28	-2.8	-2.8	0.007915452	
49		-27	-2.7	-2.7	0.010420935	
50		-26	-2.6	-2.6	0.013582969	
51		-25	-2.5	-2.5	0.0175283	
52		-24	-2.4	-2.4	0.02239453	
53		-23	-2.3	-2.3	0.020027030	
54		-22	-2.2	-2.2	0.035474593	
55		-21	-2.1	-2.1	0.043983596	
56		-20	-2	-2	0.053990967	
57		-19	-1.9	-1.9	0.065615815	
58		-18	-1.8	-1.8	0.078950158	
59		-17	-1.7	-1.7	0.094049077	
60		-16	-1.6	-1.6	0.110920835	
61		-15	-1.5	-1.5	0.129517596	
62		-14	-1.4	-1.4	0.149727466	
63		-13	-1.3	-1.3	0.171368592	
64		-12	-1.2	-1.2	0.194186055	
65		-11	-1.1	-1.1	0.217852177	
66		-10	-1	-1	0.241970725	
67		-9	-0.9	-0.9	0.26608525	
68		-8	-0.8	-0.8	0.289691553	
69		-7	-0.7	-0.7	0.312253933	
70		-6	-0.6	-0.6	0.333224603	
71		-5	-0.5	-0.5	0.352065327	
72		-4	-0.4	-0.4	0.36827014	
73		-3	-0.3	-0.3	0.381387815	
74		-2	-0.2	-0.2	0.391042694	
75		-1	-0.1	-0.1	0.396952547	
76		0	0	0	0.39894228	**The Mean**
77		1	0.1	0.1	0.396952547	

	A	B	C	D	E	F
73		-3	-0.3	-0.3	0.381387815	
74		-2	-0.2	-0.2	0.391042694	
75		-1	-0.1	-0.1	0.396952547	
76		0	0	0	0.39894228	**The Mean**
77		1	0.1	0.1	0.396952547	
78		2	0.2	0.2	0.391042694	
79		3	0.3	0.3	0.381387815	
80		4	0.4	0.4	0.36827014	
81		5	0.5	0.5	0.352065327	
82		6	0.6	0.6	0.333224603	
83		7	0.7	0.7	0.312253933	
84		8	0.8	0.8	0.289691553	
85		9	0.9	0.9	0.26608525	
86		10	1	1	0.241970725	
87		11	1.1	1.1	0.217852177	
88		12	1.2	1.2	0.194186055	
89		13	1.3	1.3	0.171368592	
90		14	1.4	1.4	0.149727466	
91		15	1.5	1.5	0.129517596	
92		16	1.6	1.6	0.110920835	
93		17	1.7	1.7	0.094049077	
94		18	1.8	1.8	0.078950158	
95		19	1.9	1.9	0.065615815	
96		20	2	2	0.053990967	
97		21	2.1	2.1	0.043983596	
98		22	2.2	2.2	0.035474593	
99		23	2.3	2.3	0.028327038	
100		24	2.4	2.4	0.02239453	
101		25	2.5	2.5	0.0175283	
102		26	2.6	2.6	0.013582969	
103		27	2.7	2.7	0.010420935	
104		28	2.8	2.8	0.007915452	
105		29	2.9	2.9	0.005952532	
106		30	3	3	0.004431848	

Normal Cumulative Distribution Function

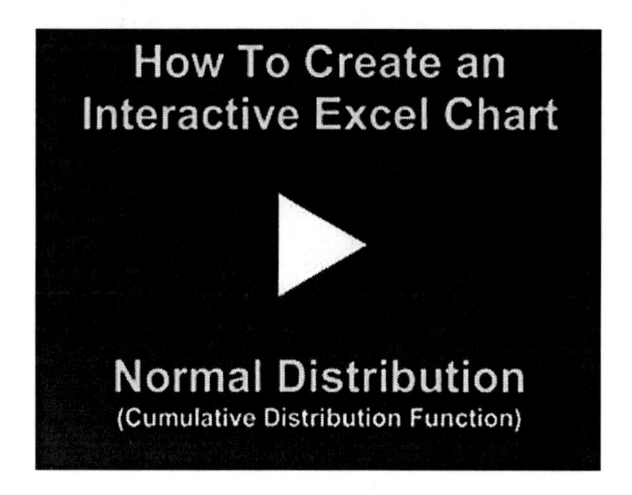

Instructional Video

Go to
http://www.youtube.com/watch?v=otTgETyzDgw
to View a
Video About How To Create
a User-Interactive Graph of the
Normal Distribution's
Cumulative Distribution Function
in Excel
(Is Your Internet Connection and Sound Turned On?)

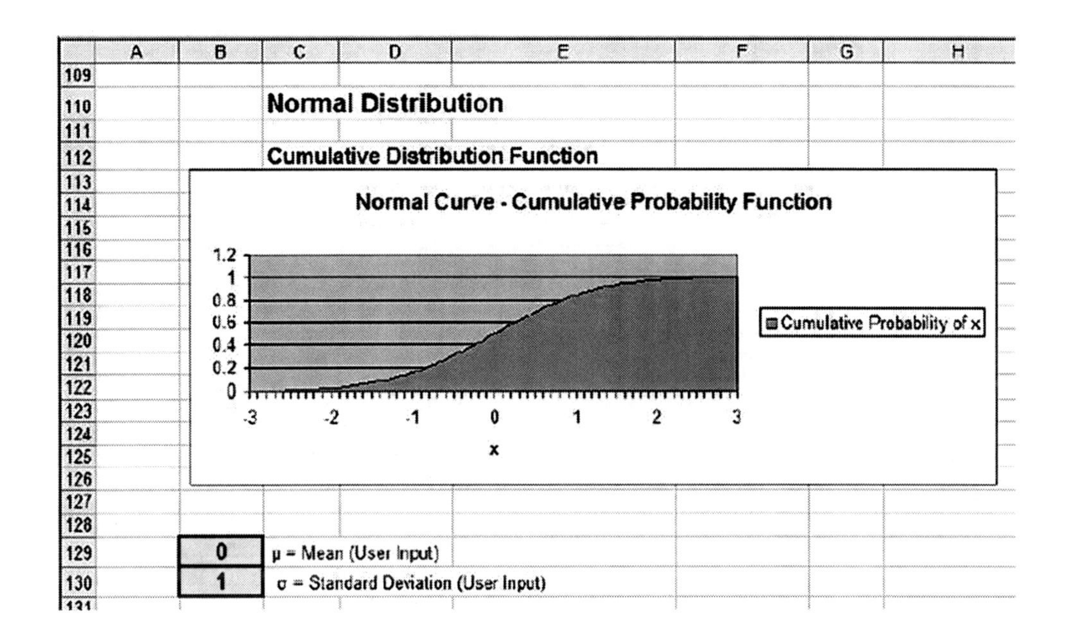
A
B
C
D
E
F
G
H
Normal Distribution
Cumulative Distribution Function
Normal Curve - Cumulative Probability Function
Cumulative Probability of x
x
0
μ = Mean (User Input)
1
σ = Standard Deviation (User Input)

	A	B	C	D	E	F
127						
128						
129		0	μ – Mean (User Input)			
130		1	σ = Standard Deviation (User Input)			
131						
132		A specific Normal Curve can be completely constructed if only the				
133		mean and standard deviation are known. These are the two				
134		user inputs for this curve. Each formula to the right of a cell that				
135		is shown below is the formula within that cell. That formula is				
136		copied to end of each column of data. The generic chart-building				
137		process is applied to this data to construct the graph below the data.				
138						
139		To create the columns of data below which are the inputs for the graph				
140		and change when new data is typed into the user inputs above, copy the				
141		given formulas below into the designated cells, then copy the formula				
142		down the column by simply dragging the formula down the column:				
143						
144		C154 =	=B154*$B$130/10			
145						
146		D154 =	=C154+$B$129			
147						
148		E154 =	=NORMDIST(D154,$B$129,$B$130,TRUE)			
149						
150						
151					**Cumulative**	
152					**Distribution**	
153				**x Axis**	**Function**	
154		-30	-3	-3	0.001349898	
155		-29	-2.9	-2.9	0.001865813	
156		28	2.8	2.8	0.00255513	

	A	B	C	D	E	F
150						
151					Cumulative	
152					Distribution	
153				x Axis	Function	
154		-30	-3	-3	0.001319898	
155		-29	-2.9	-2.9	0.001865813	
156		-28	-2.8	-2.8	0.00255513	
157		-27	-2.7	-2.7	0.003466974	
158		-26	-2.6	-2.6	0.004661188	
159		-25	-2.5	-2.5	0.006209665	
160		-24	-2.4	-2.4	0.008197536	
161		-23	-2.3	-2.3	0.01072411	
162		-22	-2.2	-2.2	0.013903448	
163		-21	-2.1	-2.1	0.017864421	
164		-20	-2	-2	0.022750132	
165		-19	-1.9	-1.9	0.02871656	
166		-18	-1.8	-1.8	0.035930319	
167		-17	-1.7	-1.7	0.044565463	
168		-16	-1.6	-1.6	0.054799292	
169		-15	-1.5	-1.5	0.066807201	
170		-14	-1.4	-1.4	0.080756659	
171		-13	-1.3	-1.3	0.096800485	
172		-12	-1.2	-1.2	0.11506967	
173		-11	-1.1	-1.1	0.135666061	
174		-10	-1	-1	0.158655254	
175		-9	-0.9	-0.9	0.184060125	
176		-8	-0.8	-0.8	0.211855399	
177		-7	-0.7	-0.7	0.241963652	
178		-6	-0.6	-0.6	0.274253118	
179		-5	-0.5	-0.5	0.308537539	
180		-4	-0.4	-0.4	0.344578258	
181		-3	-0.3	-0.3	0.382088578	
182		-2	-0.2	-0.2	0.420740291	
183		-1	-0.1	-0.1	0.460172163	
184		0	0	0	0.5	The Mean
185		1	0.1	0.1	0.539827837	

	A	B	C	D	E	F
183		-1	-0.1	-0.1	0.460172163	
184		0	0	0	0.5	The Mean
185		1	0.1	0.1	0.539827837	
186		2	0.2	0.2	0.579259709	
187		3	0.3	0.3	0.617911422	
188		4	0.4	0.4	0.655421742	
189		5	0.5	0.5	0.691462461	
190		6	0.6	0.6	0.725746882	
191		7	0.7	0.7	0.758036348	
192		8	0.8	0.8	0.788144601	
193		9	0.9	0.9	0.815939875	
194		10	1	1	0.841344746	
195		11	1.1	1.1	0.864333939	
196		12	1.2	1.2	0.88493033	
197		13	1.3	1.3	0.903199515	
198		14	1.4	1.4	0.919243341	
199		15	1.5	1.5	0.933192799	
200		16	1.6	1.6	0.945200708	
201		17	1.7	1.7	0.955434537	
202		18	1.8	1.8	0.964069681	
203		19	1.9	1.9	0.97128344	
204		20	2	2	0.977249868	
205		21	2.1	2.1	0.982135579	
206		22	2.2	2.2	0.986096552	
207		23	2.3	2.3	0.98927589	
208		24	2.4	2.4	0.991802464	
209		25	2.5	2.5	0.993790335	
210		26	2.6	2.6	0.995338812	
211		27	2.7	2.7	0.996533026	
212		28	2.8	2.8	0.99744487	
213		29	2.9	2.9	0.998134187	
214		30	3	3	0.998650102	

Normal Distribution

Graphing Outer 2% Tails
Probability Density Function

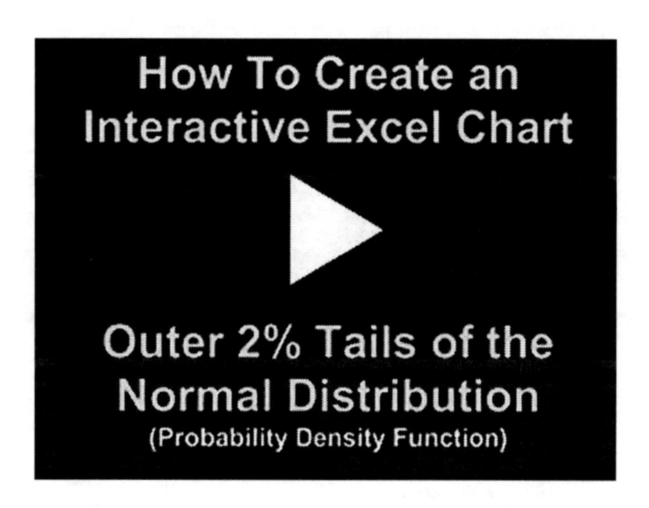

Instructional Video

Go to
http://www.youtube.com/watch?v=69-7bEIs6jQ
to View a
Video About How To Create
a User-Interactive Graph of the
Outer Tails of the
Normal Distribution's
Probability Density Function
in Excel
(Is Your Internet Connection and Sound Turned On?)

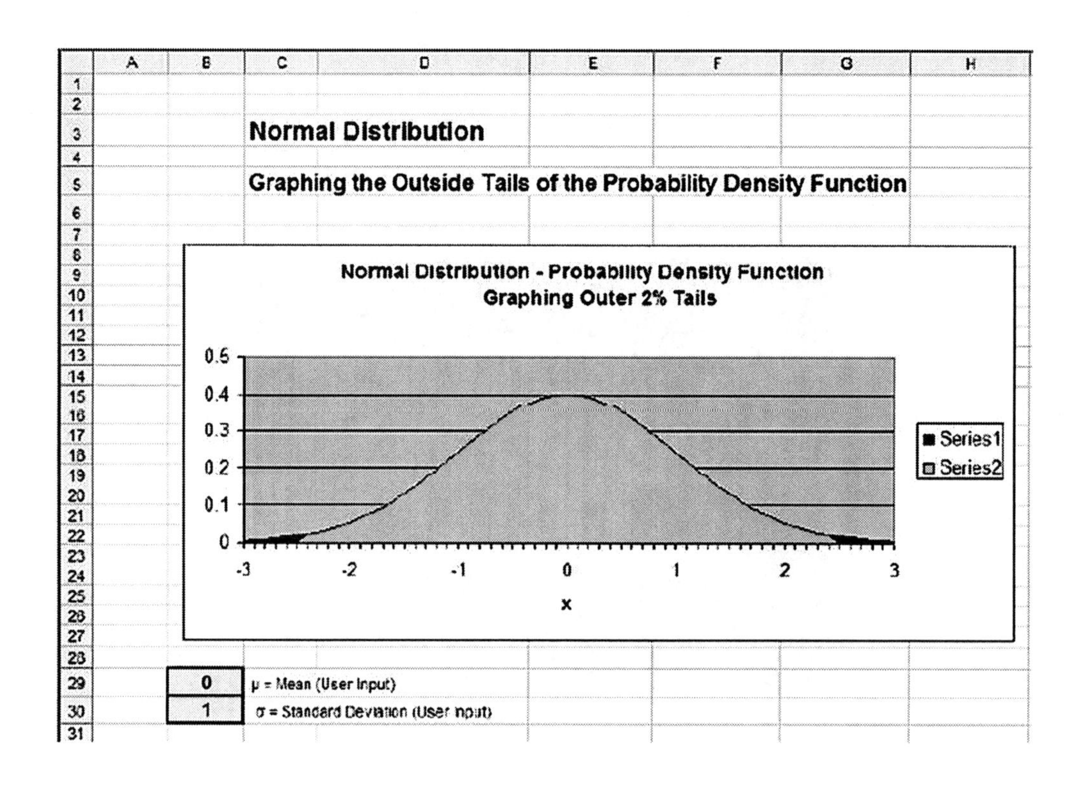
Normal Distribution
Graphing the Outside Tails of the Probability Density Function
Normal Distribution - Probability Density Function
Graphing Outer 2% Tails
Series1
Series2
x
0 μ = Mean (User Input)
1 σ = Standard Deviation (User Input)

	A	B	C	D	E	F
28						
29		0	μ = Mean (User Input)			
30		1	σ = Standard Deviation (User Input)			
31						
32						
33		Graphing outer tails simply requires creating one more identical				
34		data series of Y data. One series will have everything zeroed out				
35		except the outer tail data. The other series will have the opposite				
36		data zeroed out. Both data series need to included the data points				
37		at the boundaries. To change the colors of each graph segment				
38		Right-click on that segment and select "Format Data Series."				
39						
40		To create the columns of data below which are the inputs for the graph				
41		and change when new data is typed into the user inputs above, copy the				
42		given formulas below into the designated cells, then copy the formula				
43		down the column by simply dragging the formula down the column:				
44						
45		C56 = B56*$B$30/10				
46						
47		D56 = C56+$B$29				
48						
49		E56 = NORMDIST(D56,$B$29,$B$30,FALSE)				
50						
51		F62 = NORMDIST(D62,$B$29,$B$30,FALSE)				
52						
53						
54					Series 1	Series 2
55				x Axis	Probability y Axis	2nd Identical Probability y Axis
56		-30	-3	-3	0.004431848	0
57		-29	-2.9	-2.9	0.005952532	0
58		-28	-2.8	-2.8	0.007915452	0
59		-27	-2.7	-2.7	0.010420935	0
60		-26	-2.6	-2.6	0.013582969	0
61		-25	-2.5	-2.5	0.0175283	0
62		-24	-2.4	-2.4	0.02239453	0.02239453
63		-23	-2.3	-2.3	0	0.028327038
64		-22	-2.2	-2.2	0	0.035474593
65		-21	-2.1	-2.1	0	0.043983596

	A	B	C	D	E	F	G
52							
53							
54					Series 1	Series 2	
55				x Axis	Probability y Axis	2nd Identical Probability y Axis	
56		-30	-3	-3	0.004431848	0	
57		-29	-2.9	-2.9	0.005952532	0	
58		-28	-2.8	-2.8	0.007915452	0	
59		-27	-2.7	-2.7	0.010420935	0	
60		-26	-2.6	-2.6	0.013582969	0	
61		-25	-2.5	-2.5	0.0175283	0	
62		-24	-2.4	-2.4	0.02239453	0.02239453	
63		-23	-2.3	-2.3	0	0.028327038	
64		-22	-2.2	-2.2	0	0.035474593	
65		-21	-2.1	-2.1	0	0.043983596	
66		-20	-2	-2	0	0.053990967	
67		-19	-1.9	-1.9	0	0.065615815	
68		-18	-1.8	-1.8	0	0.078950158	
69		-17	-1.7	-1.7	0	0.094049077	
70		-16	-1.6	-1.6	0	0.110920835	
71		-15	-1.5	-1.5	0	0.129517596	
72		-14	-1.4	-1.4	0	0.149727466	
73		-13	-1.3	-1.3	0	0.171368592	
74		-12	-1.2	-1.2	0	0.194186055	
75		-11	-1.1	-1.1	0	0.217852177	
76		-10	-1	-1	0	0.241970725	
77		-9	-0.9	-0.9	0	0.26608525	
78		-8	-0.8	-0.8	0	0.289691553	
79		-7	-0.7	-0.7	0	0.312253933	
80		-6	-0.6	-0.6	0	0.333224603	
81		-5	-0.5	-0.5	0	0.352065327	
82		-4	-0.4	-0.4	0	0.36827014	
83		-3	-0.3	-0.3	0	0.381387815	
84		-2	-0.2	-0.2	0	0.391042694	
85		-1	-0.1	-0.1	0	0.396952547	
86		0	0	0	0	0.39894228	The Mean
87		1	0.1	0.1	0	0.396952547	
88		2	0.2	0.2	0	0.391042694	
89		3	0.3	0.3	0	0.381387815	

	A	B	C	D	E	F	G
82		-4	-0.4	-0.4	0	0.36827014	
83		-3	-0.3	-0.3	0	0.381387815	
84		-2	-0.2	-0.2	0	0.391042694	
85		-1	-0.1	-0.1	0	0.396952547	
86		0	0	0	0	0.39894228	The Mean
87		1	0.1	0.1	0	0.396952547	
88		2	0.2	0.2	0	0.391042694	
89		3	0.3	0.3	0	0.381387815	
90		4	0.4	0.4	0	0.36827014	
91		5	0.5	0.5	0	0.352065327	
92		6	0.6	0.6	0	0.333224603	
93		7	0.7	0.7	0	0.312253933	
94		8	0.8	0.8	0	0.289691553	
95		9	0.9	0.9	0	0.26608525	
96		10	1	1	0	0.241970725	
97		11	1.1	1.1	0	0.217852177	
98		12	1.2	1.2	0	0.194186055	
99		13	1.3	1.3	0	0.171368592	
100		14	1.4	1.4	0	0.149727466	
101		15	1.5	1.5	0	0.129517596	
102		16	1.6	1.6	0	0.110920835	
103		17	1.7	1.7	0	0.094049077	
104		18	1.8	1.8	0	0.078950158	
105		19	1.9	1.9	0	0.065615815	
106		20	2	2	0	0.053990967	
107		21	2.1	2.1	0	0.043983596	
108		22	2.2	2.2	0	0.035474593	
109		23	2.3	2.3	0	0.028327038	
110		24	2.4	2.4	0.02239453	0.02239453	
111		25	2.5	2.5	0.0175283	0	
112		26	2.6	2.6	0.013582969	0	
113		27	2.7	2.7	0.010420935	0	
114		28	2.8	2.8	0.007915452	0	
115		29	2.9	2.9	0.005952532	0	
116		30	3	3	0.004431848	0	

t Distribution

Probability Density Function

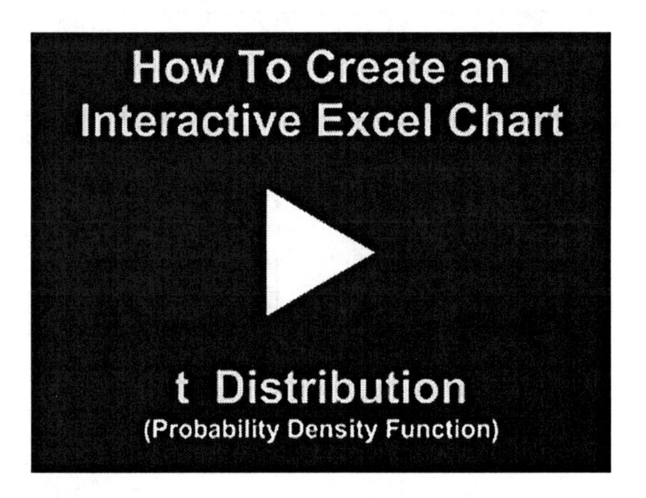

Instructional Video

Go to
http://www.youtube.com/watch?v=fqUo8-ykNbw
to View a
Video About How To Create
a User-Interactive Graph of the t-Distribution's
Probability Density Function
in Excel
(Is Your Internet Connection and Sound Turned On?)

The graphing instructions that follow for the t Distribution use the entire, ***complicated*** t Distribution formula. Excel 2003's formula for calculating the t Distribution's PDF (probability density function) produced an incorrect result. Graphing the t Distribution in Excel required the use of the actual t Distribution formula (it's not pretty, as you will see).

Excel 2010 has corrected the problem. Excel 2010's formula to calculate the PDF of the t Distribution produces a correct result.

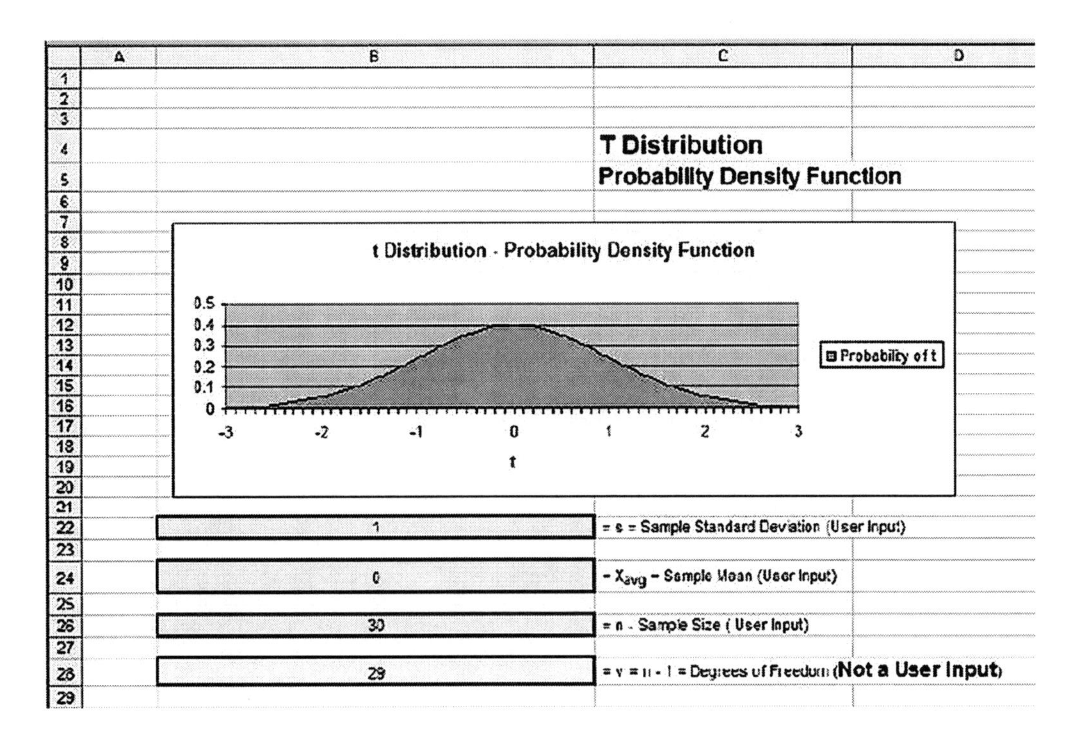

	A	B	C	D
21				
22		1	= s = Sample Standard Deviation (User Input)	
23				
24		0	= X_{avg} = Sample Mean (User Input)	
25				
26		30	= n - Sample Size (User Input)	
27				
28		29	= v = n - 1 = Degrees of Freedom (Not a User Input)	
29				
30				
31		In this case, we constructed the probability using the actual		
32		Probability Density function formula for the T Distribution.		
33		Built-in Excel functions were not used. This example illustrates		
34		how time-saving the Excel built-in functions are, as opposed		
35		to constructing the formulas by hand.		
36				
37		The reason that the actual formulas are used to graph		
38		the t distribution is that Excel's built-in function for the		
39		t Distribution Probability Density Function actually		
40		produces the wrong answer. The t Distribution reaches		
41		its highest value at approximately 0.4. The Excel TDIST		
42		function gives a value of approximately 1.0 as the highest		
43		value of the t-Distribution Probability Density function.		
44		This Excel function provides a completely incorrect		
45		answer and should not be used. Because of this,		
46		we are forced to used the actual formula to calculate		
47		the Probability Density function for the t Distribution.		
48				
49		The t Distribution is often used to analyze small samples. A		
50		t Distribution can be completely constructed if the following 3		
51		parameters are known: the sample mean (s), the sample standard		
52		deviation (σ), and the sample size (n). These are user inputs.		
53				
54		First, the t Distribution Probability Density Formula must be		
55		written in Excel. The formula of the Probability Density Function		
56		for the t Distribution is as follows:		

	A	B
54		First, the t Distribution Probability Density Formula must be
55		written in Excel. The formula of the Probability Density Function
56		for the t Distribution is as follows:
57		
58		$f(t,v) = c\,(1 + t^2/v)^{-((v+1)/2)}$
59		
60		c = [Γ((v+1)/2)] / [SQRT(v * π) * Γ(v/2)]
61		
62		= [Γ((v+1)/2)] / [SQRT(v * 3.14159265) * Γ(v/2)]
63		
64		
65		GAMMALN (x) = LN (Γ(x))
66		
67		therefore
68		
69		$\Gamma(x) = e^{LN\Gamma(x)} = e^{GAMMALNX}$ = EXP (GAMMALN(x))
70		
71		To create the columns of data below which are the inputs for the graph
72		and change when new data is typed into the user inputs above, copy the
73		given formulas below into the designated cells, then copy the formula
74		down the column by simply dragging the formula down the column, if
75		applicable.
76		
77		C92 = $B$28/2
78		
79		C94 = GAMMALN(C92)
80		
81		C98 = EXP(C94)
82		
83		C102 = ($B$28+1)/2
84		
85		C104 = GAMMALN(C102)
86		
87		C108 = EXP(C104)
88		
89		C114 = (C108)/(SQRT($B$28*3.14159265)*C98)

	A	B	C
76			
77		C92 = $B$28/2	
78			
79		C94 = GAMMALN(C92)	
80			
81		C98 = EXP(C94)	
82			
83		C102 = ($B$28+1)/2	
84			
85		C104 = GAMMALN(C102)	
86			
87		C108 = EXP(C104)	
88			
89		C114 = (C108)/(SQRT($B$28*3.14159265)*C98)	
90			
91			
92		v / 2 =	14.5
93			
94		GAMMALN (v / 2) =	23.86276584
95			
96		Γ (v / 2) = EXP (GAMMALN (v / 2)) =	
97			
98		EXP (GAMMALN (v / 2)) =	23092317917
99			
100			
101			
102		(v + 1) / 2 =	15
103			
104		GAMMALN ((v + 1) / 2) =	25.19122118
105			
106		Γ ((v + 1) / 2) = EXP (GAMMALN ((v + 1) / 2)) =	
107			
108		EXP (GAMMALN ((v + 1) / 2)) =	87178291181
109			
110			
111			
112		c = [Γ((v+1)/2)] / [SQRT(v * 3.14159265) * Γ(v/2)] =	
113			
114		=	0.395518579
115			
116		C128 = B128*$B$22/10	
117			
118		D128 = IF(C128<0,C128*(-1),C128)	

	A	B	C	D	E	F
115						
116		C128 = B128*$B$22/10				
117						
118		D128 = IF(C128<0,C128*(-1),C128)				
119						
120		E128 = $C$114*(1+(D128^2)/$B$28)^((-1)*(($B$28+1)/2))				
121		$= c\,(1 + t^2/v)^{-((v+1)/2)}$				
122						
123		F128 = C128+$B$24				
124					Probability	
125					Density	
126					Function	
127				t	f(t,v)	x Axis
128		-30	-3	3	0.006860929	-3
129		-29	-2.9	2.9	0.008676034	-2.9
130		-28	-2.8	2.8	0.010923015	-2.8
131		-27	-2.7	2.7	0.013687759	-2.7
132		-26	-2.6	2.6	0.017067669	-2.6
133		-25	-2.5	2.5	0.021171422	-2.5
134		-24	-2.4	2.4	0.026118025	-2.4
135		-23	-2.3	2.3	0.032034989	-2.3
136		-22	-2.2	2.2	0.039055441	-2.2
137		-21	-2.1	2.1	0.047314006	-2.1
138		-20	-2	2	0.05694135	-2
139		-19	-1.9	1.9	0.068057325	-1.9
140		-18	-1.8	1.8	0.080762799	-1.8
141		-17	-1.7	1.7	0.095130344	-1.7
142		-16	-1.6	1.6	0.11119415	-1.6
143		-15	-1.5	1.5	0.128939683	-1.5
144		-14	-1.4	1.4	0.148293658	-1.4
145		-13	-1.3	1.3	0.169115546	-1.3
146		-12	-1.2	1.2	0.191190864	-1.2
147		-11	-1.1	1.1	0.214227874	-1.1
148		-10	-1	1	0.23785815	-1
149		-9	-0.9	0.9	0.261641853	-0.9
150		-8	-0.8	0.8	0.285078108	-0.8
151		-7	-0.7	0.7	0.307620497	-0.7
152		-6	-0.6	0.6	0.328697264	-0.6
153		-5	-0.5	0.5	0.347735317	-0.5
154		-4	-0.4	0.4	0.364186684	-0.4
155		-3	-0.3	0.3	0.377555707	-0.3
156		-2	-0.2	0.2	0.387425032	-0.2
157		-1	-0.1	0.1	0.393478426	-0.1
158		0	0	0	0.395518579	0

	A	B	C	D	E	F	G
154		-4	-0.4	0.4	0.364186684	-0.4	
155		-3	-0.3	0.3	0.377555707	-0.3	
156		-2	-0.2	0.2	0.387425032	-0.2	
157		-1	-0.1	0.1	0.393478426	-0.1	
158		0	0	0	0.395518579	0	Mean
159		1	0.1	0.1	0.393478426	0.1	
160		2	0.2	0.2	0.387425032	0.2	
161		3	0.3	0.3	0.377555707	0.3	
162		4	0.4	0.4	0.364186684	0.4	
163		5	0.5	0.5	0.347735317	0.5	
164		6	0.6	0.6	0.328697264	0.6	
165		7	0.7	0.7	0.307620497	0.7	
166		8	0.8	0.8	0.285078108	0.8	
167		9	0.9	0.9	0.261641853	0.9	
168		10	1	1	0.23785815	1	
169		11	1.1	1.1	0.214227874	1.1	
170		12	1.2	1.2	0.191190864	1.2	
171		13	1.3	1.3	0.169115546	1.3	
172		14	1.4	1.4	0.148293658	1.4	
173		15	1.5	1.5	0.128939683	1.5	
174		16	1.6	1.6	0.11119415	1.6	
175		17	1.7	1.7	0.095130344	1.7	
176		18	1.8	1.8	0.080762799	1.8	
177		19	1.9	1.9	0.068057325	1.9	
178		20	2	2	0.05694135	2	
179		21	2.1	2.1	0.047314006	2.1	
180		22	2.2	2.2	0.039055441	2.2	
181		23	2.3	2.3	0.032034989	2.3	
182		24	2.4	2.4	0.026118025	2.4	
183		25	2.5	2.5	0.021171422	2.5	
184		26	2.6	2.6	0.017067669	2.6	
185		27	2.7	2.7	0.013687759	2.7	
186		28	2.8	2.8	0.010923015	2.8	
187		29	2.9	2.9	0.008676034	2.9	
188		30	3	3	0.006860929	3	

Binomial Distribution

Probability Density Function

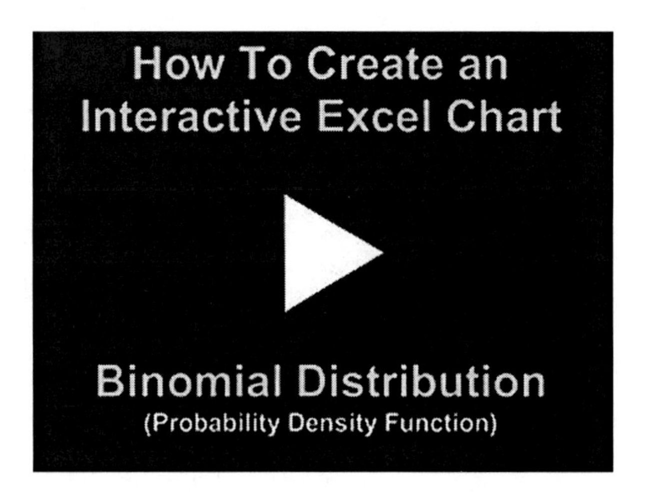

Instructional Video

Go to
http://www.youtube.com/watch?v=C8sUsVoxhZQ
to View a
Video About How To Create
a User-Interactive Graph of the
Binomial Distribution's
Probability Density Function
in Excel

(Is Your Internet Connection and Sound Turned On?)

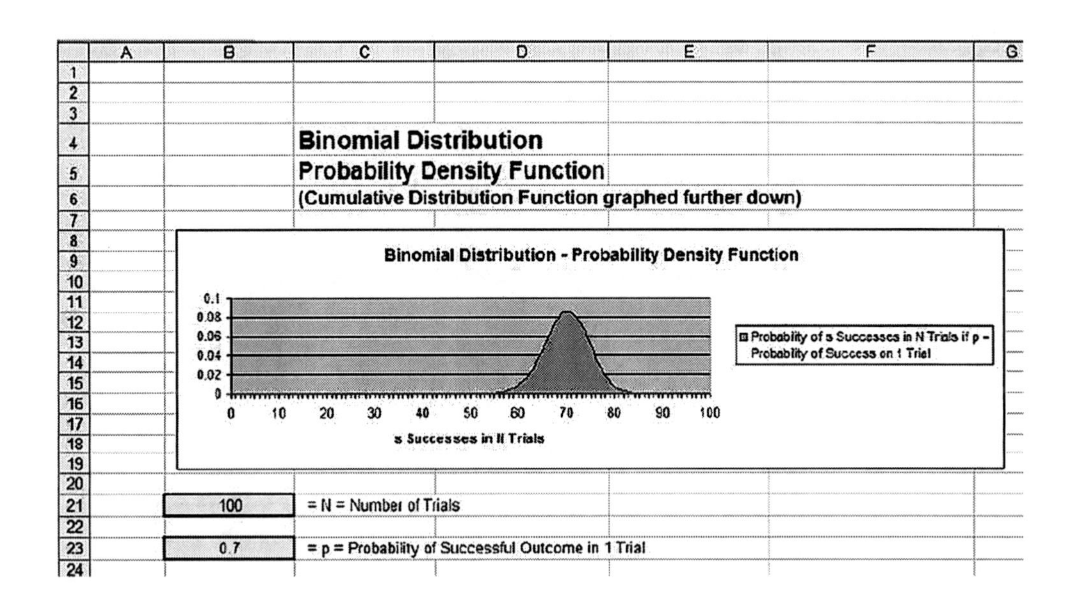
Binomial Distribution
Probability Density Function
(Cumulative Distribution Function graphed further down)
Binomial Distribution - Probability Density Function
0.1
0.08
0.06
0.04
0.02
0
0 10 20 30 40 50 60 70 80 90 100
s Successes in N Trials
Probablity of s Successes in N Trials if p = Probablity of Success on 1 Trial
100
= N = Number of Trials
0.7
= p = Probability of Successful Outcome in 1 Trial

	A	B	C	D	E
20					
21		100	= N = Number of Trials		
22					
23		0.7	= p = Probability of Successful Outcome in 1 Trial		
24					
25					
26		Each Binomial Distribution can be completely constructed if only			
27		N and p are known. The number of Successes, s, will always be a			
28		number between 0 and N.			
29					
30		The Probability Density Function calculates the probability of exactly			
31		s successes.			
32					
33		To create the columns of data below which are the inputs for the graph			
34		and change when new data is typed into the user inputs above, copy the			
35		given formulas below into the designated cells, then copy the formula			
36		down the column by simply dragging the formula down the column:			
37					
38		**D41 = IF(B41>$B$21,0,BINOMDIST(B41,$B$21,$B$23,FALSE))**			
39					
40		**S = Number Successful Trials**	**Probability Density Function**		
41		0	5.15378E-53		
42		1	1.20255E-50		
43		2	1.38894E-48		
44		3	1.05868E-46		
45		4	5.99038E-45		

	A	B	C	D	E
37					
38		D41 = IF(B41>$B$21,0,BINOMDIST(B41,$B$21,$B$23,FALSE))			
39					
40		S = Number Successful Trials	Probability Density Function		
41		0	5.15378E-53		
42		1	1.20255E-50		
43		2	1.38894E-48		
44		3	1.05868E-46		
45		4	5.99038E-45		
46		5	2.60369E-43		
47		6	9.91474E-42		
48		7	3.10662E-40		
49		8	8.42671E-39		
50		9	2.00993E-37		
51		10	4.26774E-36		
52		11	8.14751E-35		
53		12	1.40997E-33		
54		13	2.22703E-32		
55		14	3.2292E-31		
56		15	4.31995E-30		
57		16	5.35493E-29		
58		17	6.17392E-28		
59		18	6.64268E-27		
60		19	6.6893E-26		
61		20	6.32139E-25		
62		21	5.61901E-24		

	A	B	C
59		18	6.64268E-27
60		19	6.6893E-26
61		20	6.32139E-25
62		21	5.61901E-24
63		22	4.70805E-23
64		23	3.7255E-22
65		24	2.78895E-21
66		25	1.9783E-20
67		26	1.33155E-19
68		27	8.51532E-19
69		28	5.18015E-18
70		29	3.00092E-17
71		30	1.65717E-16
72		31	8.73134E-16
73		32	4.39295E-15
74		33	2.11217E-14
75		34	9.71183E-14
76		35	4.2732E-13
77		36	1.80029E-12
78		37	7.26602E-12
79		38	2.8108E-11
80		39	1.04264E-10
81		40	3.71006E-10
82		41	1.26686E-09
83		42	4.15245E-09
84		43	1.30689E-08
85		44	3.95039E-08
86		45	1.14708E-07
87		46	3.20017E-07
88		47	8.57919E-07
89		48	2.21033E-06
90		49	5.4732E-06
91		50	1.30262E-05
92		51	2.97986E-05
93		52	6.55186E-05

	A	B	C
89		48	2.21033E-06
90		49	5.4732E-06
91		50	1.30262E-05
92		51	2.97986E-05
93		52	6.55186E-05
94		53	0.000136454
95		54	0.000281182
96		55	0.000548731
97		56	0.001028871
98		57	0.001853172
99		58	0.003205774
100		59	0.005324845
101		60	0.008490169
102		61	0.012990422
103		62	0.019066588
104		63	0.026834457
105		64	0.036198564
106		65	0.046779682
107		66	0.05788395
108		67	0.068539205
109		68	0.07761057
110		69	0.083984385
111		70	0.086783865
112		71	0.085561557
113		72	0.080412019
114		73	0.071966921
115		74	0.061269135
116		75	0.049559923
117		76	0.038039414
118		77	0.027665029
119		78	0.019034486
120		79	0.0123684
121		80	0.007575645
122		81	0.004364569
123		82	0.002359706
124		83	0.001194068
125		84	0.000563866

	A	B	C
116		75	0.049559923
117		76	0.038039414
118		77	0.027665029
119		78	0.019034486
120		79	0.0123684
121		80	0.007575645
122		81	0.004364569
123		82	0.002359706
124		83	0.001194066
125		84	0.000563866
126		85	0.000247659
127		86	0.000100791
128		87	3.7845E-05
129		88	1.30451E-05
130		89	4.10406E-06
131		90	1.17042E-06
132		91	3.00107E-07
133		92	6.85027E-08
134		93	1.37497E-08
135		94	2.38912E-09
136		95	3.52081E-10
137		96	4.27877E-11
138		97	4.11703E-12
139		98	2.94073E-13
140		99	1.3862E-14
141		100	3.23448E-16
142			

Binomial Cumulative Distribution Function

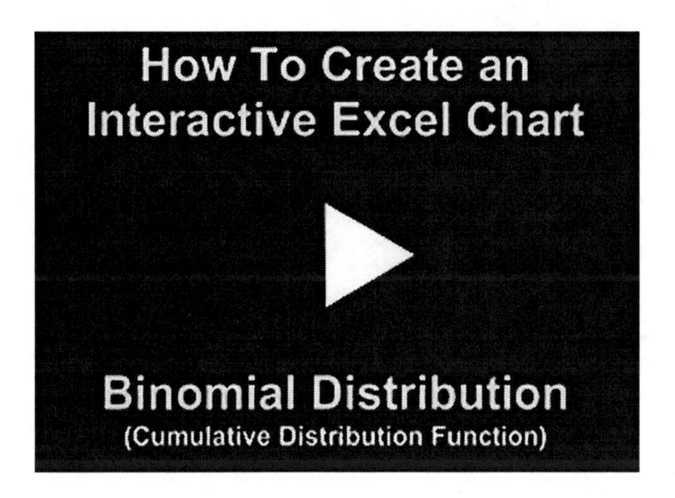

Instructional Video

Go to
http://www.youtube.com/watch?v=Ta3TTtov1WI
to View a
Video About How To Create
a User-Interactive Graph of the
Binomial Distribution's
Cumulative Distribution Function
in Excel

(Is Your Internet Connection and Sound Turned On?)

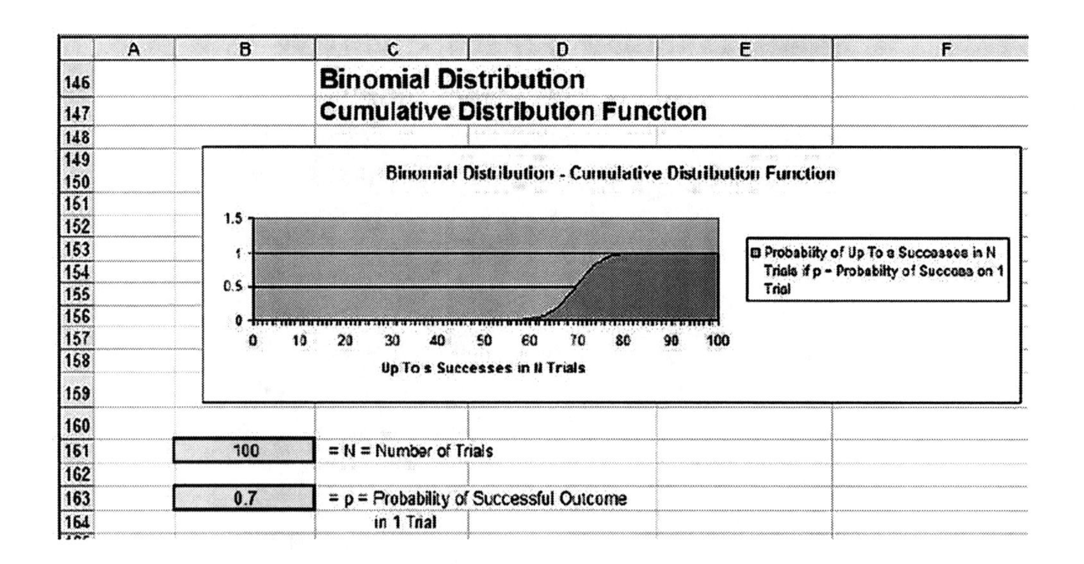
Binomial Distribution
Cumulative Distribution Function
Binomial Distribution - Cumulative Distribution Function
1.5
1
0.5
0
0 10 20 30 40 50 60 70 80 90 100
Up To s Successes in N Trials
Probability of Up To s Successes in N Trials if p = Probability of Success on 1 Trial
100
= N = Number of Trials
0.7
= p = Probability of Successful Outcome in 1 Trial

	A	B	C	D	E
20					
21		100	= N = Number of Trials		
22					
23		0.7	= p = Probability of Successful Outcome in 1 Trial		
24					
25					
26		Each Binomial Distribution can be completely constructed if only			
27		N and p are known. The number of Successes, s, will always be a			
28		number between 0 and N.			
29					
30		The Probability Density Function calculates the probability of exactly			
31		s successes.			
32					
33		To create the columns of data below which are the inputs for the graph			
34		and change when new data is typed into the user inputs above, copy the			
35		given formulas below into the designated cells, then copy the formula			
36		down the column by simply dragging the formula down the column:			
37					
38		D41 = IF(B41>$B$21,0,BINOMDIST(B41,$B$21,$B$23,FALSE))			
39					
40		S = Number Successful Trials	Probability Density Function		
41		0	5.15378E-53		
42		1	1.20255E-50		
43		2	1.38894E-48		
44		3	1.05868E-46		
45		4	5.99038E-45		

	A	B	C	D	E
178					
179		C183 = IF(B183>$B$161,1,BINOMDIST(B183,$B$161,$B$163,TRUE))			
180					
181					
182		S = Number Successful Trials	Cumulative Distribution Function		
183		0	5.15378E-53		
184		1	1.2077E-50		
185		2	1.40102E-48		
186		3	1.07269E-46		
187		4	6.09765E-45		
188		5	2.74467E-43		
189		6	1.01892E-41		
190		7	3.20851E-40		
191		8	8.74756E-39		
192		9	2.0974E-37		
193		10	4.47748E-36		
194		11	8.59526E-35		
195		12	1.49592E-33		
196		13	2.37662E-32		
197		14	3.46686E-31		
198		15	4.66663E-30		
199		16	5.8216E-29		
200		17	6.75608E-28		
201		18	7.31829E-27		
202		19	7.42113E-26		
203		20	7.0635E-25		
204		21	6.32536E-24		
205		22	5.34059E-23		
206		23	4.25956E-22		
207		24	3.21491E-21		
208		25	2.29979E-20		
209		26	1.56152E-19		
210		27	1.00768E-18		

	A	B	C
209		26	1.56152E-19
210		27	1.00768E-18
211		28	6.18783E-18
212		29	3.6197E-17
213		30	2.01914E-16
214		31	1.07505E-15
215		32	5.468E-15
216		33	2.65897E-14
217		34	1.23708E-13
218		35	5.51028E-13
219		36	2.35131E-12
220		37	9.61733E-12
221		38	3.77253E-11
222		39	1.41989E-10
223		40	5.12995E-10
224		41	1.77984E-09
225		42	5.93229E-09
226		43	1.90012E-08
227		44	5.85051E-08
228		45	1.73213E-07
229		46	4.9323E-07
230		47	1.35115E-06
231		48	3.56148E-06
232		49	9.03469E-06
233		50	2.20609E-05
234		51	5.18595E-05
235		52	0.000117370
236		53	0.000255833
237		54	0.000537015
238		55	0.001085746
239		56	0.002114617
240		57	0.003967789
241		58	0.007173563
242		59	0.012498407
243		60	0.020988576
244		61	0.033978998
245		62	0.053045586
246		63	0.079880042

	A	B	C
246		63	0.079880042
247		64	0.116078606
248		65	0.162858288
249		66	0.220742239
250		67	0.290201444
251		68	0.366892014
252		69	0.450876399
253		70	0.537660264
254		71	0.623221821
255		72	0.703633839
256		73	0.77560076
257		74	0.836869896
258		75	0.886429818
259		76	0.924469233
260		77	0.952134261
261		78	0.971168747
262		79	0.983537147
263		80	0.991112792
264		81	0.995477361
265		82	0.997837067
266		83	0.999031135
267		84	0.999595
268		85	0.999842659
269		86	0.99994345
270		87	0.999981295
271		88	0.99999434
272		89	0.999998444
273		90	0.999999615
274		91	0.999999915
275		92	0.999999983
276		93	0.999999997
277		94	1
278		95	1
279		96	1
280		97	1
281		98	1
282		99	1
283		100	1

Chi-Square Distribution

Probability Density Function

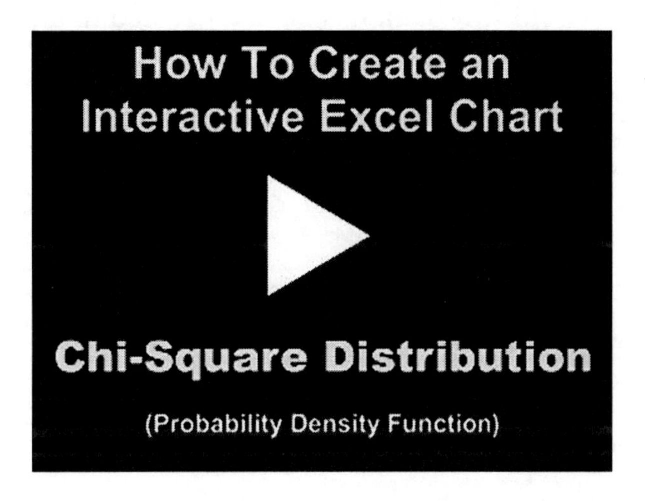

Instructional Video

Go to
http://www.youtube.com/watch?v=eDCZhK7jeYg
to View a
Video About How To Create
a User-Interactive Graph of the
Chi-Square Distribution's
Probability Density Function
in Excel

(Is Your Internet Connection and Sound Turned On?)

The actual formula is used to create the graph of the Chi-Square distribution's PDF (probability density function). Excel 2003 did not have a built-in formula that correctly calculated the Chi-Square distribution's PDF. This problem has been corrected in Excel 2010.

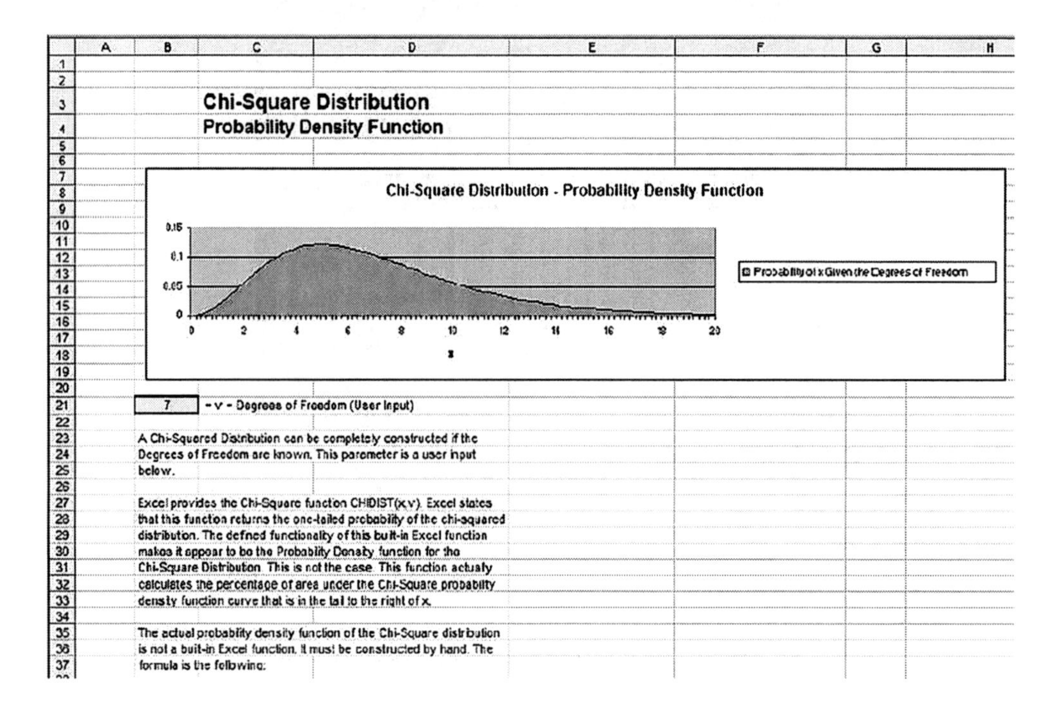

The actual probability density function of the Chi-Square distribution is not a built-in Excel function. It must be constructed by hand. The formula is the following:

Chi Square Probability Density Function =

$$= 1 / [\ (2^{v/2}) * \Gamma(v/2)\ ] * x^{(v/2-1)} * e^{-x/2}$$

To create the columns of data below which are the inputs for the graph and change when new data is typed into the user inputs above, copy the given formulas below into the designated cells, then copy the formula down the column by simply dragging the formula down the column:

C54 = CHIDIST(B54,$B$21)

D54 = (1)/((2^($B$21/2))*(EXP(GAMMALN($B$21/2))))*(B54^($B$21/2-1))*(EXP((-1)*(B54)/2))

$$= 1 / [\ (2^{v/2}) * \Gamma(v/2)\ ] * x^{(v/2-1)} * e^{-x/2}$$

x	CHIDIST(x,v) = % Curve Area Right of x	Probability Density Function
0.001	1	8.40624E-10
0.2	0.999974844	0.000430491
0.4	0.999736661	0.002203484
0.6	0.998991747	0.005494252
0.8	0.997443953	0.010205305
1	0.994828537	0.016131382
1.2	0.990926898	0.023024766
1.4	0.985671265	0.030628996

	A	B	C	D
52				
63		x	CHIDIST(x,v) = % Curve Area Right of x	Probability Density Function
54		0.001	1	8.40624E-10
55		0.2	0.999974844	0.000430491
56		0.4	0.999736561	0.002203484
57		0.6	0.998991747	0.005494252
58		0.8	0.997443953	0.010205305
59		1	0.994828537	0.016131382
60		1.2	0.990926898	0.023024766
61		1.4	0.985571265	0.030628996
62		1.6	0.978644393	0.038697524
63		1.8	0.970076446	0.047004024
64		2	0.959840369	0.055347656
65		2.2	0.947946513	0.063555272
66		2.4	0.934437081	0.07148148
67		2.6	0.919300643	0.079007607
68		2.8	0.902866967	0.086040239
69		3	0.885002234	0.092508198
70		3.2	0.865904742	0.09836092
71		3.4	0.845701103	0.103565589
72		3.6	0.824522902	0.108104803
73		3.8	0.802503859	0.111974281
74		4	0.77977741	0.115180729
75		4.2	0.756474733	0.117739889
76		4.4	0.732723089	0.11967478
77		4.6	0.70864454	0.121014129
78		4.8	0.684354954	0.121790985
79		5	0.659963234	0.122041521
80		5.2	0.635570878	0.121803989
81		5.4	0.611271555	0.121117837
82		5.6	0.587151004	0.120022958
83		5.8	0.563286964	0.118559073
84		6	0.539749359	0.116765216

	A	B	C	D
84		6	0.539749359	0.116765216
85		6.2	0.516600352	0.114679327
86		6.4	0.493894673	0.112337926
87		6.6	0.47167991	0.109775862
88		6.8	0.449996821	0.107026137
89		7	0.428879838	0.104119775
90		7.2	0.408357394	0.101085744
91		7.4	0.388452264	0.097950921
92		7.6	0.369182095	0.094740082
93		7.8	0.350559786	0.09147593
94		8	0.332593808	0.088179138
95		8.2	0.315289058	0.084868409
96		8.4	0.298646328	0.08156056
97		8.6	0.282663629	0.0782706
98		8.8	0.267336011	0.075011831
99		9	0.252656039	0.071795944
100		9.2	0.238614096	0.068633123
101		9.4	0.225198667	0.065532144
102		9.6	0.212396617	0.062500483
103		9.8	0.200193434	0.059544415
104		10	0.188573465	0.056669111
105		10.2	0.177520127	0.053878734
106		10.4	0.167016093	0.051176532
107		10.6	0.157043473	0.048564924
108		10.8	0.147583971	0.046045585
109		11	0.13861902	0.043619519
110		11.2	0.130129917	0.041287137
111		11.4	0.122097927	0.039048323
112		11.6	0.114504387	0.0369025
113		11.6	0.10733079	0.034848684
114		12	0.100558866	0.032885544
115		12.2	0.094170637	0.031011448
116		12.4	0.088148477	0.02922451
117		12.6	0.082475162	0.02752263
118		12.8	0.077133906	0.025903536
119		13	0.07210839	0.024364811
120		13.2	0.067382792	0.022903932
121		13.4	0.062941802	0.021518291

	A	B	C	D
84		6	0.539749369	0.116765216
85		6.2	0.516600352	0.114679327
86		6.4	0.493894673	0.112337926
87		6.6	0.47167991	0.109775862
88		6.8	0.449996821	0.107026137
89		7	0.428879838	0.104119776
90		7.2	0.408357394	0.101085744
91		7.4	0.388452264	0.097950921
92		7.6	0.369182095	0.094740082
93		7.8	0.350559786	0.09147593
94		8	0.332593898	0.088179138
95		8.2	0.315289058	0.084868409
96		8.4	0.298646328	0.08156056
97		8.6	0.282663629	0.0782706
98		8.8	0.267336011	0.075011831
99		9	0.252656039	0.071795944
100		9.2	0.238614096	0.068633123
101		9.4	0.225198667	0.065532144
102		9.6	0.212396617	0.062500483
103		9.8	0.200193434	0.059544415
104		10	0.188573465	0.056669111
105		10.2	0.177520127	0.053878734
106		10.4	0.167016093	0.051176532
107		10.6	0.157043473	0.048564924
108		10.8	0.147583971	0.046045585
109		11	0.13861902	0.043619519
110		11.2	0.130129917	0.041287137
111		11.4	0.122097927	0.039048323
112		11.6	0.114504387	0.0369025
113		11.8	0.10733079	0.034848684
114		12	0.100558866	0.032885544
115		12.2	0.094170637	0.031011448
116		12.4	0.088148477	0.02922451
117		12.6	0.082475162	0.02752263
118		12.8	0.077133906	0.025903536
119		13	0.07210839	0.024364811
120		13.2	0.067382792	0.022903932
121		13.4	0.062941802	0.021518291

	A	B	C	D
120		13.2	0.067382792	0.022903932
121		13.4	0.062941802	0.021518291
122		13.6	0.058770638	0.020205221
123		13.8	0.054855055	0.01896202
124		14	0.051181353	0.017785969
125		14.2	0.047736373	0.016674347
126		14.4	0.044507499	0.015624451
127		14.6	0.041482656	0.014633602
128		14.8	0.038660298	0.01369916
129		15	0.035999405	0.012818533
130		15.2	0.033519467	0.011989186
131		15.4	0.031200477	0.011208645
132		15.6	0.029032916	0.010474501
133		15.8	0.027007739	0.009784421
134		16	0.025116361	0.009136143
135		16.2	0.02335064	0.008527486
136		16.4	0.021702865	0.007956346
137		16.6	0.020165736	0.007420699
138		16.8	0.018732349	0.006918602
139		17	0.017396183	0.006448193
140		17.2	0.016151078	0.006007688
141		17.4	0.014991227	0.005595384
142		17.6	0.013911153	0.005209656
143		17.8	0.012905697	0.004848952
144		18	0.011970002	0.004511799
145		18.2	0.011099501	0.004196792
146		18.4	0.010289898	0.003902601
147		18.6	0.009537157	0.00362796
148		18.8	0.008837491	0.003371672
149		19	0.008187341	0.003132602
150		19.2	0.007583373	0.002909677
151		19.4	0.007022461	0.002701884
152		19.6	0.006501675	0.002508263
153		19.8	0.006018272	0.002327911
154		20	0.005569683	0.002159976

Poisson Distribution

Probability Density Function

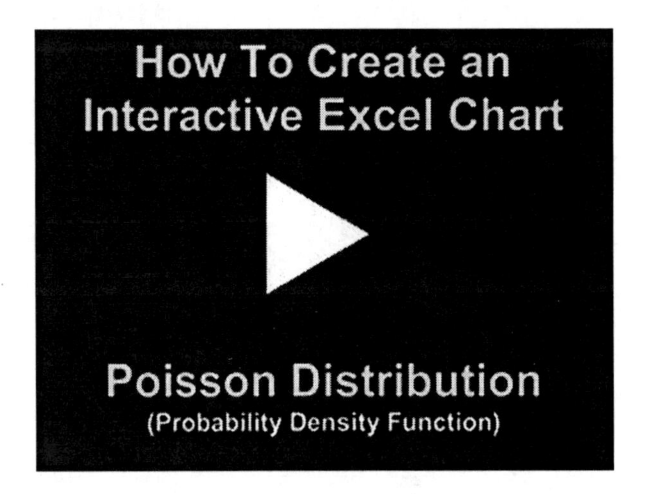

Instructional Video

Go to
http://www.youtube.com/watch?v=EoyOP5UMtL4
to View a
Video About How To Create
a User-Interactive Graph of the
Poisson Distribution's
Probability Density Function
in Excel

(Is Your Internet Connection and Sound Turned On?)

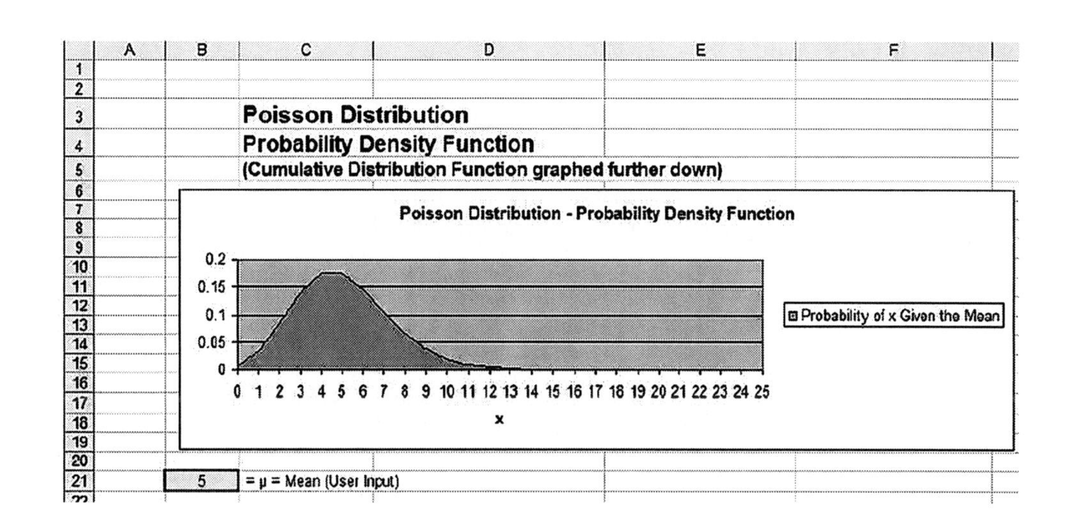
Poisson Distribution
Probability Density Function
(Cumulative Distribution Function graphed further down)
Poisson Distribution - Probability Density Function
0.2
0.15
0.1
0.05
0
0 1 2 3 4 5 6 7 8 9 10 11 12 13 14 15 16 17 18 19 20 21 22 23 24 25
x
Probability of x Given the Mean
5
= μ = Mean (User Input)

	A	B	C	D
20				
21		5	= μ = Mean (User Input)	
22				
23		The Poisson Distribution can be completely constructed		
24		if μ, the mean, is known. This is a user input.		
25				
26		The Poisson Distribution Is a discrete function so the graph		
27		has corners at each integer.		
28				
29		To create the columns of data below which are the inputs for the graph		
30		and change when new data is typed into the user inputs above, copy the		
31		given formulas below into the designated cells, then copy the formula		
32		down the column by simply dragging the formula down the column:		
33				
34		**C37 = POISSON(B37,$B$21,FALSE)**		
35				
36		**X**	**f(x,μ)**	
37		0	0.006737947	
38		1	0.033689735	
39		2	0.084224337	
40		3	0.140373896	
41		4	0.17546737	
42		5	0.17546737	
43		6	0.146222808	
44		7	[illegible]	

	A	B	C	D
33				
34		C37 = POISSON(B37,$B$21,FALSE)		
35				
36		X	f(x,µ)	
37		0	0.006737947	
38		1	0.033689735	
39		2	0.084224337	
40		3	0.140373896	
41		4	0.17546737	
42		5	0.17546737	
43		6	0.146222808	
44		7	0.104444863	
45		8	0.065278039	
46		9	0.036265577	
47		10	0.018132789	
48		11	0.008242177	
49		12	0.00343424	
50		13	0.001320862	
51		14	0.000471736	
52		15	0.000157245	
53		16	4.91392E-05	
54		17	1.44527E-05	
55		18	4.01464E-06	
56		19	1.05648E-06	
57		20	2.64121E-07	
58		21	6.2886E-08	
59		22	1.42923E-08	
60		23	3.10701E-09	
61		24	6.47295E-10	
62		25	1.29459E-10	

Poisson Cumulative Distribution Function

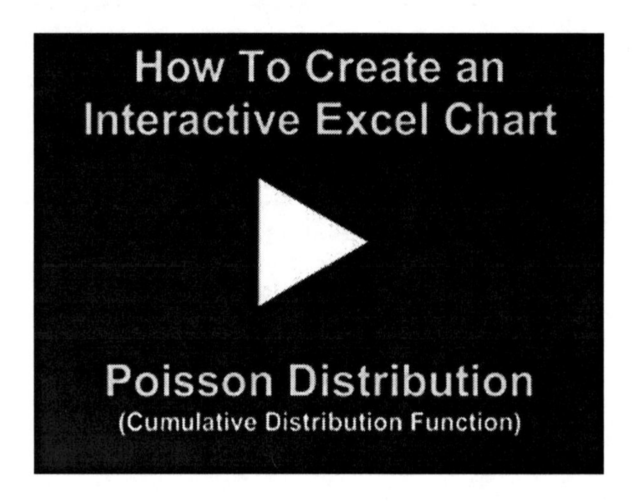

Instructional Video

Go to
http://www.youtube.com/watch?v=Fu95WzmKqn8
to View a
Video About How To Create
a User-Interactive Graph of the
Poisson Distribution's
Cumulative Distribution Function
in Excel

(Is Your Internet Connection and Sound Turned On?)

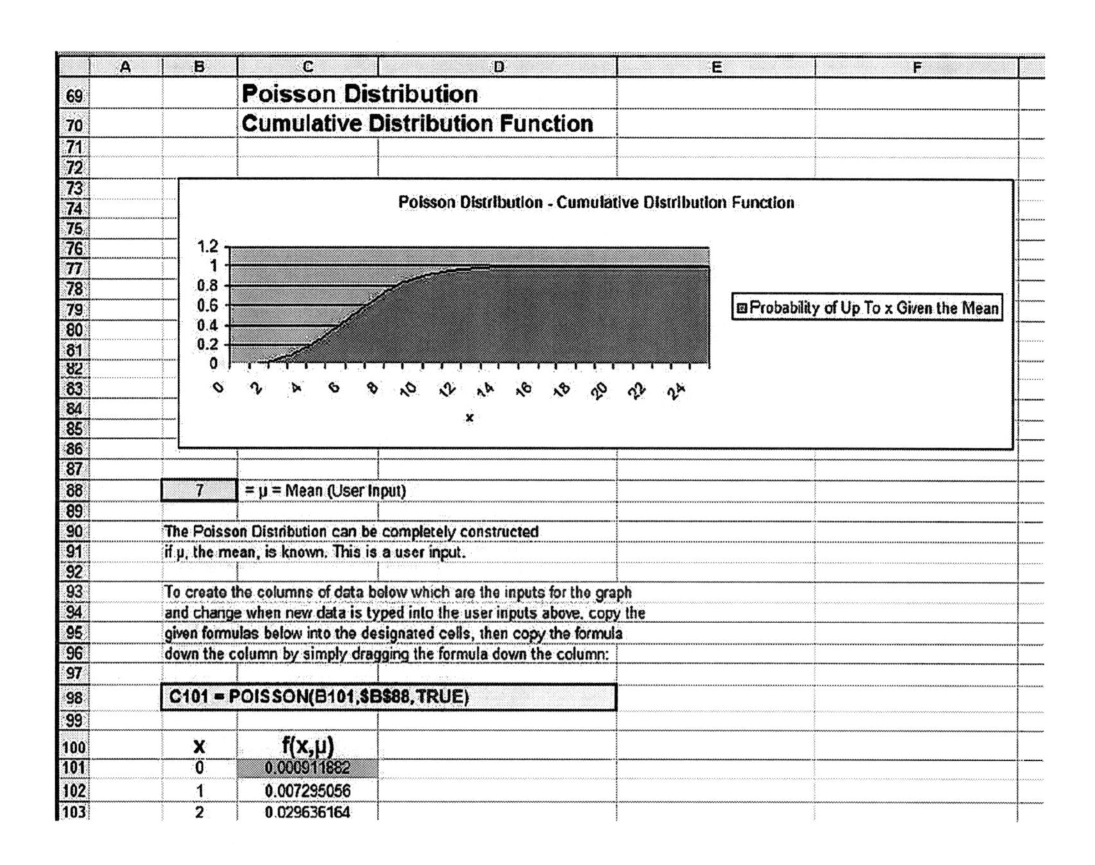

Poisson Distribution
Cumulative Distribution Function
Poisson Distribution - Cumulative Distribution Function
Probability of Up To x Given the Mean
x
7
= μ = Mean (User Input)
The Poisson Distribution can be completely constructed
if μ, the mean, is known. This is a user input.
To create the columns of data below which are the inputs for the graph
and change when new data is typed into the user inputs above, copy the
given formulas below into the designated cells, then copy the formula
down the column by simply dragging the formula down the column:
C101 = POISSON(B101,$B$88,TRUE)
x
f(x,μ)
0
0.000911882
1
0.007295056
2
0.029636164

	A	B	C	D
97				
98		C101 = POISSON(B101,$B$88,TRUE)		
99				
100		X	f(X,µ)	
101		0	0.000911882	
102		1	0.007295056	
103		2	0.029636164	
104		3	0.081765416	
105		4	0.172991608	
106		5	0.300700276	
107		6	0.449711056	
108		7	0.598713836	
109		8	0.729091268	
110		9	0.830495937	
111		10	0.901479206	
112		11	0.946650377	
113		12	0.973000227	
114		13	0.987188607	
115		14	0.994282798	
116		15	0.99759342	
117		16	0.999041817	
118		17	0.999638216	
119		18	0.999870149	
120		19	0.999955598	
121		20	0.999985505	
122		21	0.999995474	
123		22	0.999998646	
124		23	0.999999611	
125		24	0.999999893	
126		25	0.999999971	

Weibull Distribution

Probability Density Function and Cumulative Distribution Function

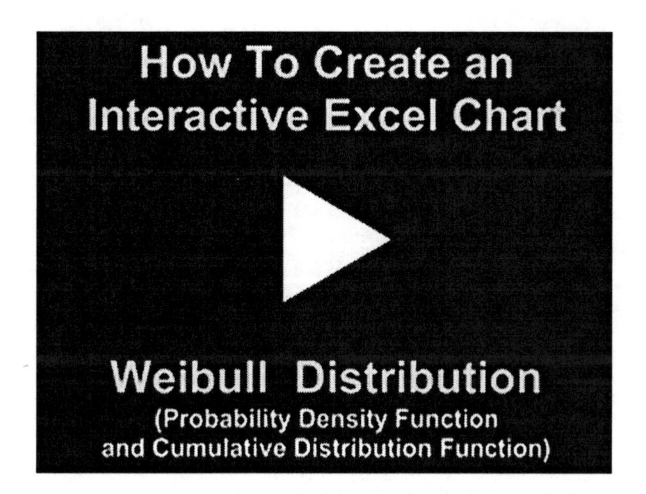

Instructional Video

Go to
http://www.youtube.com/watch?v=Ln1J9u5duis
to View a
Video About How To Create
a User-Interactive Graph of the
Weibull Distribution's
Probability Density Function and
Cumulative Distribution Function
in Excel

(Is Your Internet Connection and Sound Turned On?)

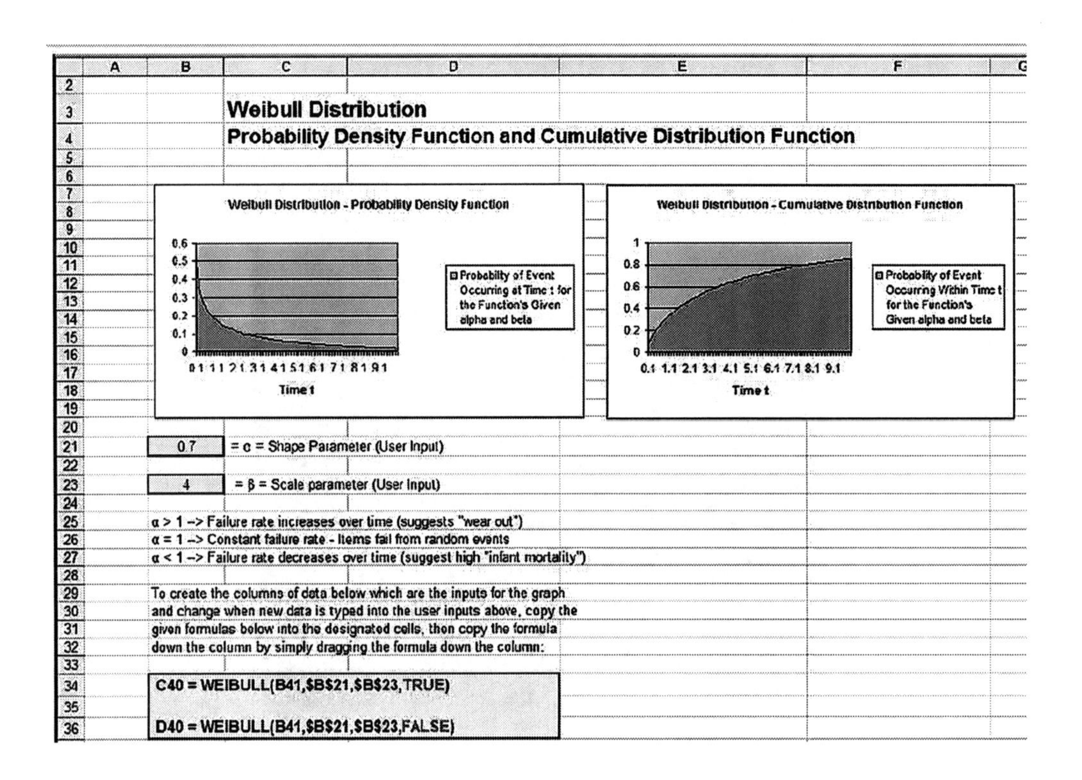
Weibull Distribution
Probability Density Function and Cumulative Distribution Function
Weibull Distribution - Probability Density Function
Probability of Event Occurring at Time t for the Function's Given alpha and beta
Time t
Weibull Distribution - Cumulative Distribution Function
Probability of Event Occurring Within Time t for the Function's Given alpha and beta
Time t
0.7 = α = Shape Parameter (User Input)
4 = β = Scale parameter (User Input)
α > 1 --> Failure rate increases over time (suggests "wear out")
α = 1 --> Constant failure rate - Items fail from random events
α < 1 --> Failure rate decreases over time (suggest high "infant mortality")
To create the columns of data below which are the inputs for the graph and change when new data is typed into the user inputs above, copy the given formulas below into the designated cells, then copy the formula down the column by simply dragging the formula down the column:
C40 = WEIBULL(B41,$B$21,$B$23,TRUE)
D40 = WEIBULL(B41,$B$21,$B$23,FALSE)

	A	B	C	D
33				
34		C40 = WEIBULL(B41,$B$21,$B$23,TRUE)		
35				
36		D40 = WEIBULL(B41,$B$21,$B$23,FALSE)		
37				
38				
39		t	Cumulative Distribution Function	Probability Density Function
40		0.1	0.072818837	0.490705184
41		0.2	0.115579634	0.380194458
42		0.3	0.150522138	0.323349158
43		0.4	0.180881266	0.28601243
44		0.5	0.208050972	0.2586201
45		0.6	0.23280248	0.237201735
46		0.7	0.255622506	0.219745438
47		0.8	0.276844728	0.205097613
48		0.9	0.296711384	0.192538188
49		1	0.31540588	0.181588864
50		1.1	0.333071621	0.171916451
51		1.2	0.349823634	0.163279947
52		1.3	0.365756104	0.155499622
53		1.4	0.380947482	0.148438012
54		1.5	0.395464059	0.141987723
55		1.6	0.409362538	0.136063313
56		1.7	0.422691936	0.130595735
57		1.8	0.435495017	0.125528423
58		1.9	0.447809391	0.120814478
59		2	0.459668375	0.116414596
60		2.1	0.471101665	0.112295527
61		2.2	0.48213589	0.108428901
62		2.3	0.492795044	0.104790324
63		2.4	0.50310085	0.101358676
64		2.5	0.51307306	0.098115551
65		2.6	0.522729704	0.095044819
66		2.7	0.532087295	0.092132266

	A	B	C	D
66		2.7	0.532087295	0.092132266
67		2.8	0.541161009	0.089365307
68		2.9	0.549964835	0.086732746
69		3	0.558511704	0.084224582
70		3.1	0.566813598	0.081831847
71		3.2	0.574881651	0.079546468
72		3.3	0.582726227	0.077361156
73		3.4	0.590356997	0.075269306
74		3.5	0.597783003	0.073264915
75		3.6	0.605012712	0.071342615
76		3.7	0.612054071	0.069497107
77		3.8	0.618914547	0.067724115
78		3.9	0.625601168	0.066019336
79		4	0.632120559	0.064378902
80		4.1	0.638478974	0.062799249
81		4.2	0.644682325	0.061277081
82		4.3	0.650736205	0.059809349
83		4.4	0.656645914	0.058393225
84		4.5	0.662416482	0.057026081
85		4.6	0.668052682	0.055705472
86		4.7	0.673559052	0.054429121
87		4.8	0.67893991	0.0531949
88		4.9	0.68419937	0.052000823
89		5	0.689341361	0.050845028
90		5.1	0.694369593	0.049725774
91		5.2	0.699287669	0.048641424
92		5.3	0.70409899	0.047590441
93		5.4	0.708806821	0.046571381
94		5.5	0.713414285	0.045582882
95		5.6	0.717924373	0.044623661
96		5.7	0.722339953	0.043692509
97		5.8	0.726663772	0.042788281
98		5.9	0.73089847	0.041909898
99		6	0.735046579	0.041056337
100		6.1	0.739110533	0.040226629
101		6.2	0.74309267	0.039419856
102		6.3	0.74699524	0.038635147

	A	B	C	D
102		6.3	0.74699524	0.038635147
103		6.4	0.750820407	0.037871676
104		6.5	0.754570256	0.037128656
105		6.6	0.758246795	0.036405341
106		6.7	0.761851958	0.03570102
107		6.8	0.76538761	0.035015015
108		6.9	0.76885555	0.034346682
109		7	0.772257515	0.033695406
110		7.1	0.77559518	0.033060599
111		7.2	0.778870165	0.032441703
112		7.3	0.782084033	0.031838181
113		7.4	0.785238296	0.031249521
114		7.5	0.788334416	0.030675236
115		7.6	0.791373807	0.030114857
116		7.7	0.794357836	0.029567935
117		7.8	0.797287828	0.029034041
118		7.9	0.800165065	0.028512765
119		8	0.802990789	0.028003711
120		8.1	0.805766202	0.027506501
121		8.2	0.808492471	0.027020771
122		8.3	0.811170727	0.026546174
123		8.4	0.813802066	0.026082374
124		8.5	0.816387551	0.025629049
125		8.6	0.818926215	0.02518589
126		8.7	0.821425058	0.024752598
127		8.8	0.823879054	0.024328888
128		8.9	0.826291146	0.023914483
129		9	0.828662252	0.023509119
130		9.1	0.830993262	0.023112538
131		9.2	0.833285044	0.022724496
132		9.3	0.835538438	0.022344754
133		9.4	0.837754264	0.021973082
134		9.5	0.839933316	0.021609261
135		9.6	0.84207637	0.021253076
136		9.7	0.844184179	0.020904322
137		9.8	0.846257476	0.020562799
138		9.9	0.848296974	0.020228316
139		10	0.850303367	0.019900687

Exponential Distribution

Probability Density Function and Cumulative Distribution Function

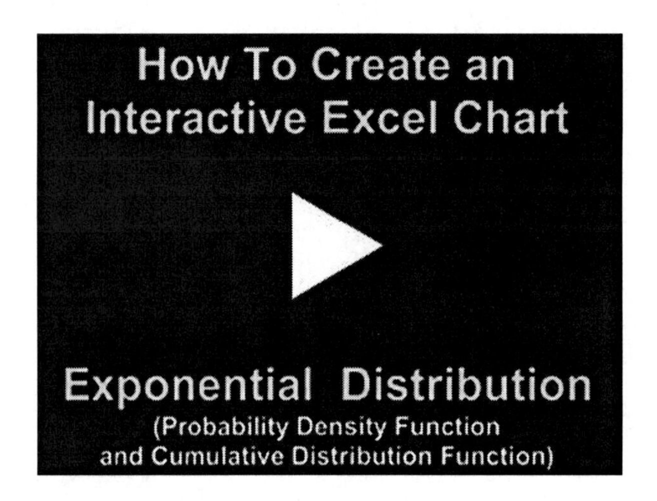

Instructional Video

Go to
http://www.youtube.com/watch?v=yyWP6OImzxM
to View a
Video About How To Create
a User-Interactive Graph of the
Exponential Distribution's
Probability Density Function and
Cumulative Distribution Function
in Excel

(Is Your Internet Connection and Sound Turned On?)

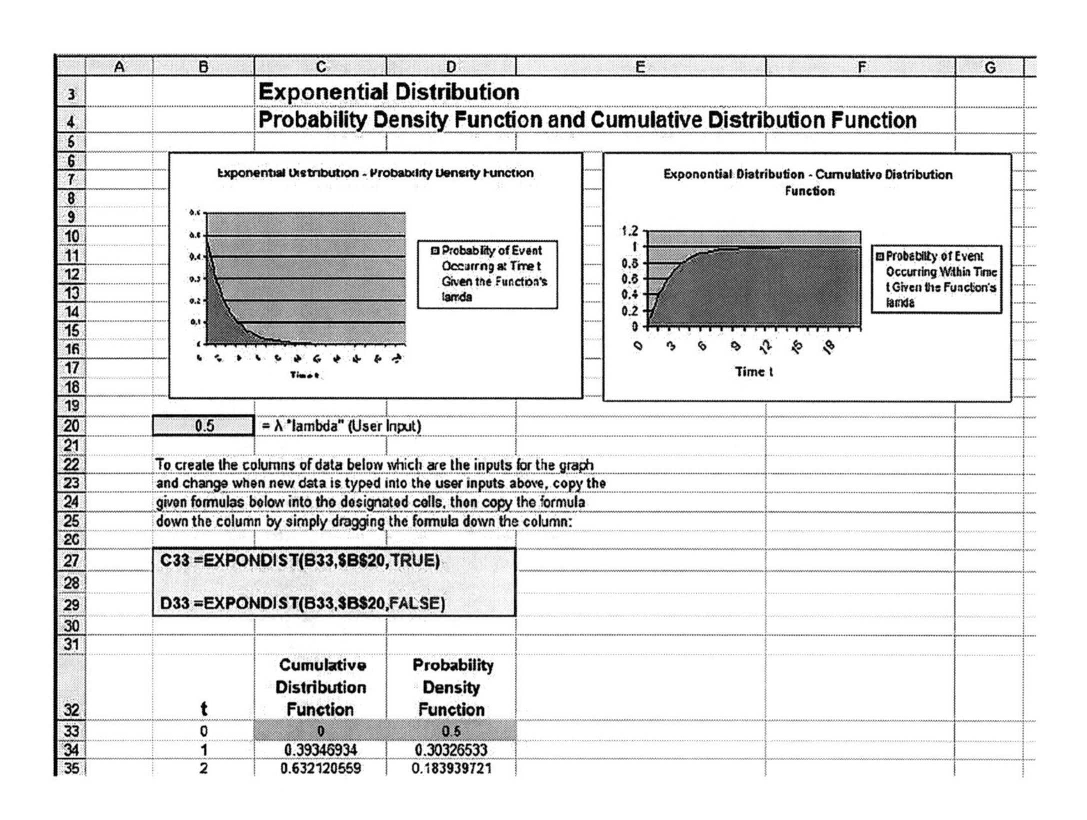
Exponential Distribution
Probability Density Function and Cumulative Distribution Function
Exponential Distribution - Probability Density Function
Probability of Event Occurring at Time t Given the Function's lamda
Exponential Distribution - Cumulative Distribution Function
1.2
1
0.8
0.6
0.4
0.2
0
0
3
6
9
12
15
18
Time t
Probability of Event Occurring Within Time t Given the Function's lamda
0.5
= λ "lambda" (User Input)
To create the columns of data below which are the inputs for the graph and change when new data is typed into the user inputs above, copy the given formulas below into the designated cells, then copy the formula down the column by simply dragging the formula down the column:
C33 =EXPONDIST(B33,$B$20,TRUE)
D33 =EXPONDIST(B33,$B$20,FALSE)
t
Cumulative Distribution Function
Probability Density Function
0
0
0.5
1
0.39346934
0.30326533
2
0.632120559
0.183939721

	A	B	C	D
26				
27		C33 =EXPONDIST(B33.$B$20.TRUE)		
28				
29		D33 =EXPONDIST(B33,$B$20,FALSE)		
30				
31				
32		t	Cumulative Distribution Function	Probability Density Function
33		0	0	0.5
34		1	0.39346934	0.30326533
35		2	0.632120559	0.183939721
36		3	0.77686984	0.11156508
37		4	0.864664717	0.067667642
38		5	0.917915001	0.041042499
39		6	0.950212932	0.024893534
40		7	0.969802617	0.015098692
41		8	0.981684361	0.009157819
42		9	0.988891003	0.005554498
43		10	0.993262053	0.003368973
44		11	0.995913229	0.002043386
45		12	0.997521248	0.001239376
46		13	0.998496561	0.00075172
47		14	0.999088118	0.000455941
48		15	0.999446916	0.000276542
49		16	0.999664537	0.000167731
50		17	0.999796532	0.000101734
51		18	0.99987659	6.17049E-05
52		19	0.999925148	3.74269E-05
53		20	0.9999546	2.27E-05
54				

Hypergeometric Distribution

Probability Density Function

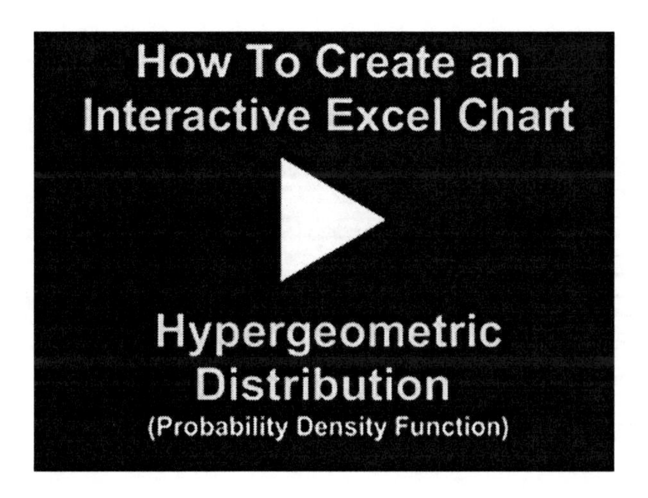

Instructional Video

Go to
http://www.youtube.com/watch?v=uYG_oXWMqyo
to View a
Video About How To Create
a User-Interactive Graph of the
Hypergeometric Distribution's
Probability Density Function
in Excel

(Is Your Internet Connection and Sound Turned On?)

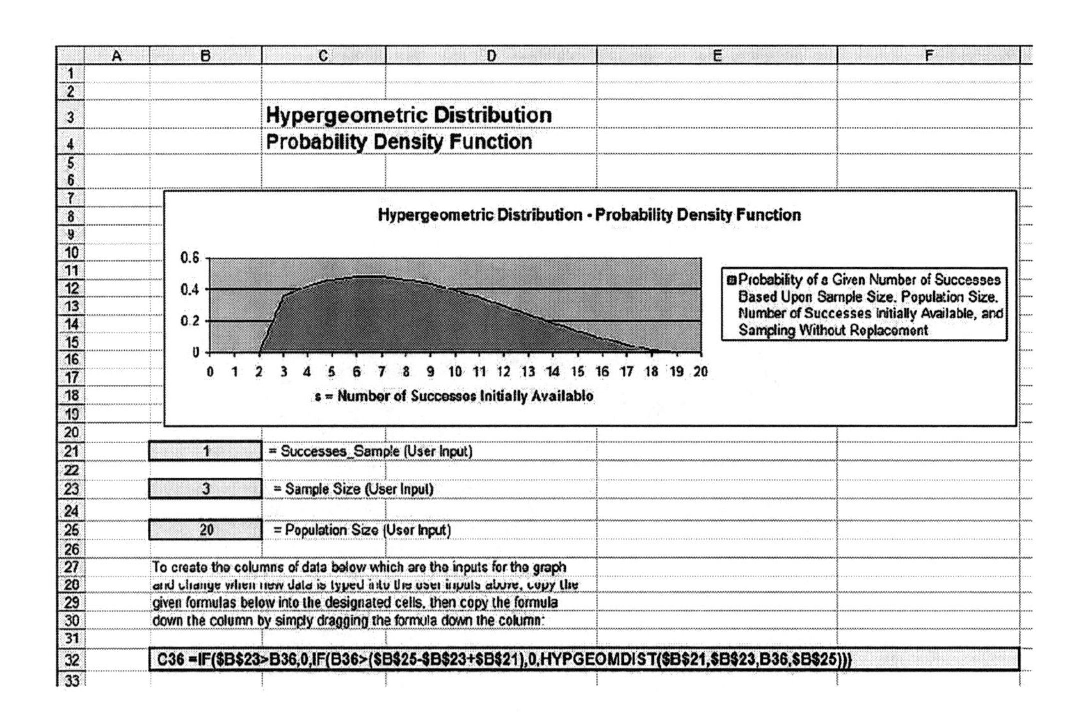
Hypergeometric Distribution
Probability Density Function
Hypergeometric Distribution - Probability Density Function
0.6
0.4
0.2
0
0 1 2 3 4 5 6 7 8 9 10 11 12 13 14 15 16 17 18 19 20
s = Number of Successes Initially Available
Probability of a Given Number of Successes Based Upon Sample Size, Population Size, Number of Successes Initially Available, and Sampling Without Replacement
1 = Successes_Sample (User Input)
3 = Sample Size (User Input)
20 = Population Size (User Input)
To create the columns of data below which are the inputs for the graph and change when new data is typed into the user inputs above, copy the given formulas below into the designated cells, then copy the formula down the column by simply dragging the formula down the column:
C36 =IF($B$23>B36,0,IF(B36>($B$25-$B$23+$B$21),0,HYPGEOMDIST($B$21,$B$23,B36,$B$25)))

	A	B	C	D	E	F
20						
21		1	= Successes Sample (User Input)			
22						
23		3	= Sample Size (User Input)			
24						
25		20	= Population Size (User Input)			
26						
27		To create the columns of data below which are the inputs for the graph				
28		and change when new data is typed into the user inputs above, copy the				
29		given formulas below into the designated cells, then copy the formula				
30		down the column by simply dragging the formula down the column:				
31						
32		C36 =IF($B$23>B36,0,IF(B36>($B$25-$B$23+$B$21),0,HYPGEOMDIST($B$21,$B$23,B36,$B$25)))				
33						
34						
35		s – Number of Successes Initially Available	Probability Density Function			
36		0	0			
37		1	0			
38		2	0			
39		3	0.357894737			
40		4	0.421052632			
41		5	0.460526316			
42		6	0.478947368			
43		7	0.478947368			
44		8	0.463157895			
45		9	0.434210526			
46		10	0.394736842			
47		11	0.347368421			
48		12	0.294736842			
49		13	0.239473684			
50		14	0.184210526			
51		15	0.131578947			
52		16	0.084210526			
53		17	0.044736842			

	A	B	C
34			
35		s = Number of Successes Initially Available	Probability Density Function
36		0	0
37		1	0
38		2	0
39		3	0.357894737
40		4	0.421052632
41		5	0.460526316
42		6	0.478947368
43		7	0.478947368
44		8	0.463157895
45		9	0.434210526
46		10	0.394736842
47		11	0.347368421
48		12	0.294736842
49		13	0.239473684
50		14	0.184210526
51		15	0.131578947
52		16	0.084210526
53		17	0.044736842
54		18	0.015789474
55		19	0
56		20	0
57			

Correctable Reasons Why Your Data Is Not Normally Distributed

In the ideal world, all of your data samples are normally distributed. In this case you can usually apply the well-known parametric statistical tests such as ANOVA, the t Test, and regression to the sampled data.

What can you do if your data does not appear to be normally distributed?

You can either:

- Apply nonparametric tests to the data. Nonparametric tests do not rely on the underlying data to have any specific distribution

- Evaluate whether your "non-normal" data was really normally-distributed before it was affected by one of the seven correctable causes listed below:

The Biggest 7 Correctable Causes of Non-Normality in Data Samples

1) Outliers

Too many outliers can easily skew normally-distributed data. If you can identify and remove outliers that are caused by error in measurement or data entry, you might be able to obtain normally-distributed data from your skewed data set. Outliers should only be removed if a specific cause of their extreme value is identified. The nature of the normal distribution is that some outliers will occur.

Outliers should be examined carefully if there are more than would be expected.

2) Data has been affected by more than one process

It is very important to understand all of the factors that can affect data sample measurement. Variations to process inputs might skew what would otherwise be normally-distributed output data. Input variation might be caused by factors such as shift changes, operator changes, or frequent changes in the underlying process. A common symptom that the output is being affected by more than one process is the occurrence of more than one mode (most commonly occurring value) in the output. In such a situation, you must isolate each input variation that is affecting the output. You must then isolate the overall effect which that variation had on the output. Finally, you must remove that input variation's effect from output measurement. You may find that you now have normally-distributed data.

3) Not enough data

A normal process will not look normal at all until enough samples have been collected. It is often stated that 30 is the where a "large" sample starts. If you have collected 50 or fewer samples and do not have a normally-distributed sample, collect at least 100 samples before re-evaluating the normality of the population from which the samples are drawn.

4) Measuring devices that have poor resolution

Devices with poor resolution may round off incorrectly or make continuous data appear discrete. You can, of course, use a more accurate measuring device. A simpler solution is to use a much larger sample size to smooth out sharp edges.

5) A different distribution describes the data

Some forms of data inherently follow different distributions. For example, radioactive decay is described by the exponential distribution.

The Poisson distribution describes events event that tend to occur at predictable intervals over time, such as calls over a switchboard, number of defects, or demand for services. The lengths of time between occurrences of Poisson-distributed processes are described by the exponential distribution. The uniform distribution describes events that have an equal probability of occurring. Application of the Gamma distribution often based on intervals between Poisson-distributed events, such as queuing models and the flow of items through a manufacturing process. The Beta distribution is often used for modeling planning and control systems such are PERT and CPM. The Weibull distribution is used extensively to model time between failure of manufactured items, finance, and climatology. It is important to become familiar with the applications of other distributions. If you know that the data is described by a different distribution than the normal distribution, you will have to apply the techniques of that distribution or use nonparametric analysis techniques.

6) Data approaching zero or a natural limit

If the data has a large number of value than are near zero or a natural limit, the data may appear to be skewed. In this case, you may have to adjust all data by adding a specific value to all data being analyzed. You need to make sure that all data being analyzed is "raised" to the same extent.

7) Only a subset of process' output is being analyzed

If you are sampling only a specific subset of the total output of a process, you are likely not collecting a representative sample from the process and therefore will not have normally distributed samples. For example, if you are evaluating manufacturing samples that occur between 4 and 6AM and not an entire shift, you might not obtain the normally-distributed sample that a whole shift would provide. It is important to ensure that your sample is representative of an entire process.

If you are unable to obtain a normally-distributed data sample, you can usually apply non-parametric tests to the data.

Statistical Mistakes To Avoid

1) Assuming that correlation equals causation

This is, of course, not true. However, if you find a correlation, you should look hard for links between the two. The correlation may be pure chance, but then again, it may not be. A correlation is a reason to look for underlying causes behind the behavior. Correlation is often a symptom of a larger issue.

2) Not graphing and eyeballing the data prior to performing regression analysis

Always graph the data before you do regression analysis. You'll know immediately whether you're dealing with linear regression, non-linear regression, or completely unrelated data that can't be regressed.

3) Not doing correlation analysis on all variables prior to performing regression

You'll save yourself a lot of time if you can remove any input variables that have a low correlation with the dependent (output –> Y) variable or that have a high correlation with another input variable (this error is called multicollinearity). In the 2nd case, you would want to remove the input variable from the highly correlated pair of input variables that has the lowest correlation with the output variable.

4) Adding a large number of new input variables into a regression analysis all at once

You always want to add new input variables one at a time and run a separate regression each time a new input variable is added. The changes you observe to the output of the regression will tell you whether the new input variable adds to the predictive power of the regression equation. Adjusted r squared only increases when a new variable adds greater predictive power to the regression equation.

5) Applying input variables to a regression equation that are outside of the value of the original input variables that were used to create the regression equation

Here is an example to illustrate why this might produce totally invalid results. Suppose that you created a regression equation that predicted a child's weight based upon the child's age, and then you provided an adult age as an input. This regression equation would predict a completely incorrect weight for the adult, because adult data was not used to construct the original regression equation.

6) Not examining the residuals in regression

You should always at least eyeball the residuals. If the residuals show a pattern, your regression equation is not explaining all of the behavior of the data.

7) Only evaluating r square in a regression equation

In the output of regression performed in Excel, there are actually four very important components of the output that should be looked at. The chapter in this manual about regression goes into deep detail about how to analyze regression output.

8) Not drawing a representative sample from a population

This is usually solved by taking a larger sample and using a random sampling technique such as nth-ing (sampling every nth object in the population).

9) Drawing a conclusion without applying the proper statistical analysis

This occurs quite often when people simply eyeball the results instead of performing a hypothesis test to determine if the observed change has at least an 80% chance (or whatever level of certainty you desire) of not being pure chance.

10) Drawing a conclusion before a statistically significant result has been reached

This is often caused by choosing a statistical test requiring a lot of samples but depending on a low sample rate. A common occurrence of this would be performing multivariate testing on a web site that does not have sufficient traffic. Such a test is likely to be concluded prematurely. A better solution might be to perform a number of successive A/B split-tests in place of multivariate analysis. You get a lot more testing done a lot faster, and correctly.

11) Analyzing non-normal data with the normal distribution

Data should always be eyeballed in a Histogram and then and analyzed for normality before using the normal distribution-based testing techniques. If the data is not normally distributed, you must use data fitting techniques to determine which statistical distribution most closely fits the data.

12) Not removing outliers prior to statistical analysis

A couple of outliers can skew results badly. Once again, eyeball the data and determine what belongs and what doesn't. Another solution is to test the data using nonparametric analysis. Many nonparametric tests are not significantly affected by a few outliers.

13) Not controlling or taking into account other variables besides the one(s) being testing when using the t test, ANOVA, or hypothesis tests.

Other variables that not part of the test need to be held as constant as possible during the above tests or your answer might be invalid without you realizing it.

14) Using the wrong t test

The t-test to be applied depends upon factors such as whether or samples have the same size and variance. It is important to pick the right t-test before starting. Read the t-Test chapters in this e-manual.

15) Attempting to apply the wrong type of hypothesis test

There are 4 ways that the data must be classified before the correct hypothesis test can be selected. Chapters 8 and 9 of this manual cover hypothesis testing in detail.

16) Not using Excel

This point may sound a little self-serving, but knowing how to do this stuff in Excel is a real time-saver, particularly if you are in marketing, and especially if you're an Internet marketer. You'll never need to pick up another thick confusing statistics text book or figure how to work those confusing statistics tables ever again. I've actually thrown out all of my statistics text books (well, not quite, I sold them on eBay).

17) Always requiring 95% certainty

This could really slow you down. For example, if I'm A/B split-testing keywords or ads in an AdWords campaign, I will typically pick a winner when my split-tester tells me that it is 80% sure that one result is better than the other. Achieving 95% certainty would often take too long.

18) Thinking it is impossible to get a statistically significant sample if your target market is large

The sample size you need from a large population is probably quite a bit smaller than you think. Nationwide surveys are normally within a percentage point or two from real answer after only several thousand interviews have been conducted. That of course depends hugely on obtaining a representative sample to interview.

19) Not taking steps to ensure that your sample is normally distributed when analyzing with the normal distribution

One way to ensure that you have a normally distributed sample for analysis is to take a large number of samples (at 30) with each sample consisting of several random and simultaneously-chosen data points and then take the mean from each sample. Make that mean the sample. Your samples will now be Normally distributed. You can take as few as 2 data points per sample, but the more data points per sample, the fewer data points it will take for your samples (each sample is the mean of the data points collected for one sample) to appear to be Normally distributed. If you are taking only 3 data points per sample, you may have to collect over samples (that would be a total of 300 data points) for your samples to appear to be Normally distributed.

A statistical theory called the Central Limit Theory states that the means of samples (at least 30 samples and each sample having at least 2 data points that are averaged to get a mean, which will be the value of the sample) will be Normally distributed, no matter how the underlying population is distributed. You can then perform statistical analysis on that group using the normal distribution-based techniques.

20) Using covariance analysis instead of correlation analysis

The output of covariance analysis is dependent upon the scale used to measure the data. Different scales of measurement can produce completely different results on the same data if covariance analysis is used.

Correlation analysis is completely independent of the scale used to measure the data. Different scales of measurement will produce the same results on a data set using correlation analysis, unlike covariance analysis.

21) Using a one-tailed test instead of a two-tailed test when accuracy is needed

If accuracy it what you are seeking, it might be better to use the two-tailed when performing, for example, a hypothesis test. The two-tailed test is more stringent than the one-tailed test because the outer regions (I refer to them the regions of uncertainty) are half the size in a two-tailed test than in a one-tailed test. The two-tailed test tells you merely that the means are different. The one-tailed test tells you that the means are different in one specific direction.

22) Not using nonparametric tests when analyzing small samples of unknown distribution

The t Distribution should only be used in small sample analysis if the population from which the samples were drawn was Normally distributed. Nonparametric tests are valid when the population distribution is not known, or is known not to be Normally distributed. Using the t distribution in either of these cases for small sample analysis is invalid.

There is, however, a clever way around the normality requirement. If you are able to take a large number of representative samples from the population of unknown distribution, arrange the samples into random groups of equal size. Each group should have the same number of samples. Calculate the average for each group. These averages will be normally distributed and you can perform all statistical analysis mentioned in this manual on those group averages.

Check it out in Excel. Use the Excel random number generator to generate 1,000 random numbers between 0 and 1. Split those 1,000 numbers into 200 random groups of 5 each. Take the average of each group. Create a histogram in Excel of those group averages. The histogram shape will be that of the normal curve. The group averages will be normally distributed.

The Central Limit Theorem predicts this. The group averages will be normally distributed no matter what distribution the underlying population from which the samples were drawn had. You can now perform all statistical analysis that you read about in this manual on those group averages. Amazing huh?

Meet Mark the Author

Mark Harmon is a master number cruncher. Creating overloaded Excel spreadsheets loaded with complicated statistical analysis is his idea of a good time. His profession as an Internet marketing manager provides him with the opportunity and the need to perform plenty of meaningful statistical analysis.

Mark Harmon is also a natural teacher. As an adjunct professor, he spent five years teaching more than thirty semester-long courses in marketing and finance at the Anglo-American College in Prague, Czech Republic and the International University in Vienna, Austria. During that five-year time period, he also worked as an independent marketing consultant in Czechoslovakia and then the Czech Republic and performed long-term assignments for more than one hundred clients. His years of teaching and consulting have honed his ability to present difficult subject matter in an easy-to-understand way.

Mark Harmon received a degree in electrical engineering from Villanova University and MBA in marketing from the Wharton School.

If You Liked
Practical and Clear Graduate Statistics in Excel

You'll Love
Step-By-Step Optimization With Excel Solver

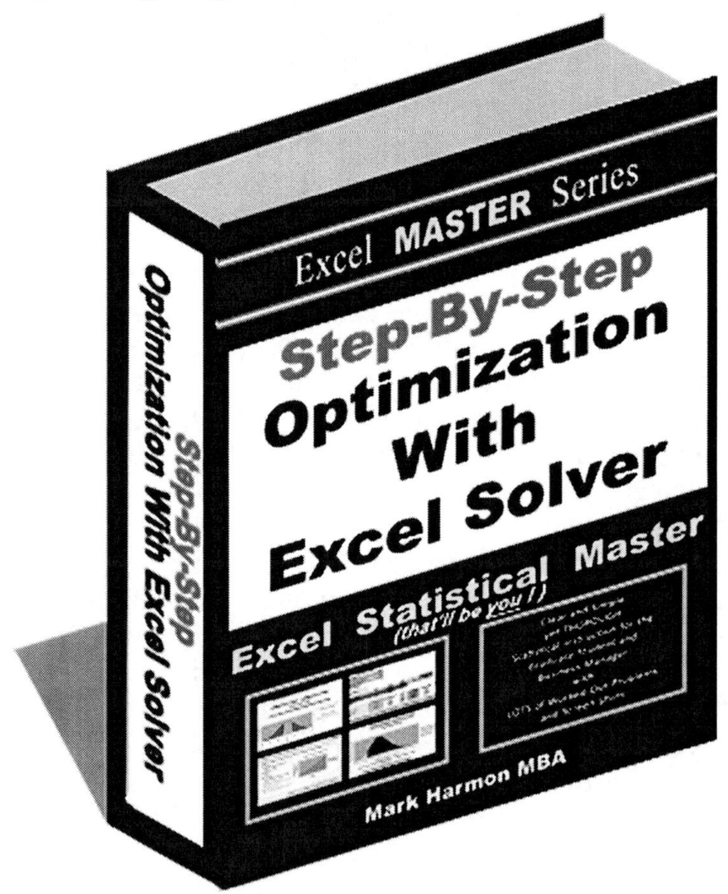

Find Out What's In It On The Next Page

If You Wish To Purchase It, Go To:

http://www.amazon.com/Step---Step-Optimization-Solver-ebook/dp/B005O2FOFE

Here's What's Inside Step-By-Step Optimization With Excel Solver

For anyone who wants to be operating at a high level with the Excel Solver quickly, this is the book for you. Step-By-Step Optimization With Excel Solver is more than 200+ pages of simple yet thorough explanations on how to use the Excel Solver to solve today's most widely known optimization problems. Loaded with screen shots that are coupled with easy-to-follow instructions, this book will simplify many difficult optimization problems and make you a master of the Excel Solver almost immediately.

Here are just some of the Solver optimization problems that are solved completely with simple-to-understand instructions and screen shots in this book:

• The famous "Traveling Salesman" problem using Solver's Alldifferent constraint and the Solver's Evolutionary method to find the shortest path to reach all customers. This also provides an advanced use of the Excel INDEX function.

• The well-known "Knapsack Problem" which shows how optimize the use of limited space while satisfying numerous other criteria.

• How to perform nonlinear regression and curve-fitting on the Solver using the Solver's GRG Nonlinear solving method.

• How to solve the "Cutting Stock Problem" faced by many manufacturing companies who are trying to determine the optimal way to cut sheets of material to minimize waste while satisfying customer orders.

• Portfolio optimization to maximize return or minimize risk.

• Venture capital investment selection using the Solver's Binary constraint to maximize Net Present Value of selected cash flows at year

• Clever use of the If-Then-Else statements makes this a simple problem.

• How use Solver to minimize the total cost of purchasing and shipping goods from multiple suppliers to multiple locations.

• How to optimize the selection of different production machine to minimize cost while fulfilling an order.

• How to optimally allocate a marketing budget to generate the greatest reach and frequency or number of inbound leads at the lowest cost.

Step-By-Step Optimization With Excel Solver has complete instructions and numerous tips on every aspect of operating the Excel Solver. You'll fully understand the reports and know exactly how to tweek all of the Solver's settings for total custom use. The book also provides lots of inside advice and guidance on setting up the model in Excel so that it will be as simple and intuitive as possible to work with.

All of the optimization problems in this book are solved step-by-step using a 6-step process that works every time. In addition to detailed screen shots and easy-to-follow explanations on how to solve every optimization problem in the book, a link is provided to download an Excel workbook that has all problems completed exactly as they are in this book.

Step-By-Step Optimization With Excel Solver is exactly the book you need if you want to be optimizing at an advanced level with the Excel Solver quickly.

Find Out What Readers Are Saying About Step-By-Step Optimization With Excel Solver On The Next Page

If You Wish To Purchase It, Go To:
http://www.amazon.com/Step---Step-Optimization-Solver-ebook/dp/B005O2FOFE

Here's What Readers Are Saying About Step-By-Step Optimization With Excel Solver

"I do strategic planning for large enterprise IT projects. While I was familiar with Excel Solver from my MBA class, this book opened a world of ideas to me. The examples are fantastic and gave me great ideas into how set up my problems into solver.

For me, the best part of the book is the insight it provides into reading and translating the reports that solver generates. When I am presenting my solutions to senior management and I am asked what if something changed, I can very quickly answer based on a sensitivity analysis and my interpretation of the report.

Overall a great book that is very easy to read, understand and implement. Thanks Mr. Harmon, keep up the good work."

Michael Langebeck
Morrisville, North Carolina

"I'm finished with school (Financial Economics major) and currently work for a fortune 400 company as a business analyst. I find that the statistics and optimization manuals are indispensable reference tools throughout the day.

I keep both eManuals loaded on my ipad at all times just in case I have to recall a concept I don't use all the time. Its easier to recall the concepts from the eManuals rather then trying to sift through the convoluted banter in a text book, and for that I applaud the author!

In a business world where I need on demand answers now this optimization eManual is the perfect tool.

I just recently used the bond investment optimization problem to build a model in excel and help my VP understand that a certain process we're doing wasn't maximizing our resources.

That's the great thing about this manual, you can use any practice problem (with a little outside thinking) to mold it into your own real life problem and come up with answers that matter in the work place!!"

Sean Ralston
Sr. Financial Analyst
Enogex LLC
Oklahoma City, Oklahoma

"Step-By-Step Optimization With Excel Solver is the "Missing Manual" for the Excel Solver. It is pretty difficult to find good documentation anywhere on solving optimization problems with the Excel Solver. This book came through like a champ!

Optimization with the Solver is definitely not intuitive, but this book is. I found it very easy to work through every single one of the examples. The screen shots are clear and the steps are presented logically. The downloadable Excel spreadsheet with all example completed was quite helpful as well.

Once again, it really amazing how little understandable documentation there is on doing real-life optimization problems with Solver.

For example, just try to find anything anywhere about the well-known Traveling Salesman Problem (a salesman needs to find the shortest route to visit all customers once). It is a tricky problem for sure, but this book showed a quick and easy way to get it done. I'm not sure I would have ever figured that problem out, or some the other problems in the book, without this manual.

I can say that this is the book for anyone who wants or needs to get up to speed on an advanced level quickly with the Excel Solver. It appears that every single aspect of using the Solver seems to be covered thoroughly and yet simply. <u>The author presents a lot of tricks in how to set the correct Solver settings to get it to do exactly what you want.</u>

<u>The book flows logically. It's an easy read. Step-By-Step Optimization With Excel Solver got me up to speed on the Solver quickly and without to much mental strain at all. I can definitely recommend this book.</u>"

Pam Copus
Sonic Media Inc.

"As Graduate student of the Graduate Program in International Studies (GPIS) at Old Dominium University, I'm required to have a thorough knowledge of Excel in order to use it as a tool for interpreting data, conducting research and analysis.

I've always found the Excel Solver to be one of the more difficult Excel tools to totally master. Not any more. <u>This book was so clearly written that I was able to do almost every one of advanced optimization examples in the book as soon as I read through it once.</u>

<u>I can tell that the author really made an effort to make this manual as intuitive as possible. The screen shots were totally clear and logically presented.</u>

Some of the examples that were very advanced, such as the venture capital investment example, had screen shot after screen shot to ensure clarity of the difficult Excel spreadsheet and Solver dialogue boxes.

It definitely was "Step-By-Step" just like the title says. I must say that I did have to cheat a little bit and look at the Excel spreadsheet with all of the book's example that is downloadable from the book. The spreadsheet was also a great help.

Step-By-Step Optimization With Excel Solver is not only totally easy to understand and follow, but it is also very complete. I feel like I'm a master of the Solver. I have purchased a couple of other books in the Excel MaSter Series (the Excel Statistical Master and the Advanced Regression in Excel book) and they have all been excellent.

I am lucky to have come across this book because the graduate program that I am in has a number of optimization assignments using the Solver. Thanks Mark for such an easy-to-follow and complete book on using the Solver. It really saved me a lot of time in figuring this stuff out."

Federico Catapano
Graduate Student
International Studies Major
Old Dominion University
Norfolk, Virginia

"Excel Solver is a tool that most folks never use. I was one of those people. I was working on a project, and was told that solver might be helpful. I did some research online, and was more confused than ever. I started looking for a book that might help me. I got this book, and was not sure what to expect.

It surpassed my expectations! The book explains the concepts behind the solver, the best way to set up the "problem", and how to use the tool effectively. It also gives many examples including the files. The files are stored online, and you can download them so you can see everything in excel.

The author does a fantastic job on this book. While I'm not a solver "expert", I am definitely much smarter about it than I was before. ***Trust me, if you need to understand the solver tool, this book will get you there! "***

Scott Kinsey
Missouri

"The author, Mark, has a writing style that is easy to follow, simple, understandable, with clear examples that are easy to follow. This book is no exception.

Mark explains how solver works, the different types of solutions that can be obtained and when to use one or another, explains the content and meaning of the reports available. Then he presents several examples, goes about defining each problem, setting it up in excel and in solver and interpreting the solution.

It is a really good book that teaches you how to apply solver (linear programming) to a problem.

Luis R. Heimpel
El Paso, Texas

If You Wish To Purchase It, Go To:
http://www.amazon.com/Step---Step-Optimization-Solver-ebook/dp/B005O2F0FE

Download

All Excel Spreadsheets

Used In This Manual Right Here

Go to

http://excelmasterseries.com/Excel_Statistical_Master_All_Spreadsheets.xls

CPSIA information can be obtained at www.ICGtesting.com
Printed in the USA
LVOW031051220412

278511LV00032B/2/P